An Introduction to Practical Biochemistry

THIRD EDITION

An Introduction to
Practical Biochemistry

THIRD EDITION

David T. Plummer

Senior Lecturer in Biochemistry
King's College
University of London

McGRAW-HILL BOOK COMPANY

London · New York · St Louis · San Francisco · Auckland · Bogotá
Guatemala · Hamburg · Lisbon · Madrid · Mexico · Montreal
New Delhi · Panama · Paris · San Juan · São Paulo · Singapore
Sydney · Tokyo · Toronto

Published by
McGRAW-HILL Book Company (UK) Limited
MAIDENHEAD · BERKSHIRE · ENGLAND

British Library Cataloguing in Publication Data
Plummer, David T.
   An introduction to practical biochemistry.
   ——3rd ed.
   1. Biological chemistry——Technique
   I. Title
   574.19′2′028   QP519.7

   ISBN 0-07-084165-9

Library of Congress Cataloging-in-Publication Data
Plummer, David T.
   An introduction to practical biochemistry.

   Bibliography: p.
   Includes index.
   1. Biological chemistry——Laboratory manuals.
I. Title.
QP519.P58   1987   574.19′2   87-4186
ISBN 0-07-084165-9

Cover photograph
   "Dr A Lesk, Laboratory of Molecular
   Biology/Science Photo Library"
   (Molecular computer graphic of the DNA
   molecule, showing its double helix
   structure.)

234   BP   898
Typeset by Advanced Filmsetters (Glasgow) Ltd
and printed and bound in Great Britain by The Bath Press, Bath

To Ruth, Malcolm, Jonathan and Martyn

Proverbs IV v. 7 (A.V.) Wisdom is the principal thing; therefore get wisdom: and with all thy getting get understanding

Contents

Preface to the third edition

An Introduction to Practical Biochemistry was first published in 1971 and the second edition appeared in 1978 so the time is now ripe for a revised and updated version of the book. The broad plan remains essentially the same as the earlier editions, starting with practical exercises on techniques then moving on to experiments with biological molecules and finally a section on biological structure and metabolism. However, there have been a number of changes within this framework although the third edition has retained nine of the ten chapters used in the text of 1978. These chapters have been revised and rewritten with the chapter on separation methods divided in the third edition into chromatography and electrophoresis and that on lipids and membranes also divided into the two separate topics. A new section on fluorescence has also been included in the chapter on spectrophotometry. The remaining five chapters on safety in laboratories, immunochemical methods, photosynthesis and respiration, cell fractionation and molecular biology are completely new. In spite of the increase in the number of chapters from 10 to 16, the size of the book remains essentially the same. This has been achieved by removing some experiments and limiting the number of practical exercises in each chapter. Further savings have been made by condensing the amount of material devoted to the theory but this has not been removed completely otherwise the book would become merely a laboratory cook book. The theoretical introduction, together with the further reading list for each chapter, should give the student a solid foundation for understanding the practical aspects of biochemistry.

Today most undergraduates and technicians in the biological and life sciences require a working knowledge of biochemistry and especially the practical aspects of the subject. The book should therefore be suitable for many BSc and technical courses in biochemistry by careful selection from the wide range of experiments provided. It could also be used on MSc courses and be a helpful source of reference for research workers who need to understand and use biochemical techniques in their work.

I must of course thank all the past undergraduates of Chelsea College without whom this work would not have been possible. I am also grateful to our technicians who prepared the classes and particularly Mr Ted Crutcher and Mr Ned Ashcroft who have made several valuable suggestions. I wish to thank all those colleagues who have provided experiments for inclusion in the book including Dr Peter Butterworth, Dr Alan Ebringer, Dr Derek Evered, Dr Geoff Hall and Dr Mike Perry of King's College, London and Mr R. Young of Newcastle Polytechnic.

Finally I must thank my present and past research assistants and particularly Dr David Obatomi, Mr Taiwo Fashola and Miss Mojgan Hossein-Nia who have read the manuscript and the proofs. Last but not least I must thank my wife Ruth for her love and encouragement and my family for their patience during the preparation of the manuscript.

David Plummer

Technical notes

Solutions

A list of solutions required to carry out an investigation is given for each experiment together with the quantities needed. The figure 10 or 100 printed on the same line as the heading 'materials' means that the weights and volumes have been calculated assuming 10 or 100 students working in pairs. The amount required for a given class can, therefore, be readily calculated from this information. In arriving at the quantities needed, it has been assumed that estimations will be carried out in duplicate. In addition, an extra allowance over and above the bare minimum has been added and the solutions made up to the next most convenient volume. For example, if 300 ml is required for an experiment, then the recommended volume is given as 500 ml, but if 400 ml is needed then the volume suggested is 1 litre.

Experiments with 100 marked by the 'materials' can be carried out by a class of that size but where 10 is marked there is probably some limitation on the availability of equipment or expensive reagents.

Finally, these quantities assume that the reagents are divided equally among the class with a small quantity in reserve. If reagents are left on a bench and people help themselves then much bigger quantities will need to be made up.

SECTION ONE Laboratory work

1. Safety in laboratories

General safety measures

Laboratory safety may appear at first sight to be rather a dull subject and the temptation may be to read this section only superficially or not at all. However, the view of the subject changes rapidly if you find yourself in the middle of a fire or the victim of an accident and by this time ignorance can be dangerous or even fatal.

Laboratory safety equipment

Laboratories can be dangerous places in which to work and all users need to be aware of the potential hazards and to know what to do in cases of emergency. When starting work in a new laboratory, it is important to become familiar with the layout of the room and the location of the safety equipment. The position of the emergency exits, fire alarm and extinguishers should be known so that appropriate action can be taken in the event of fire. It is also important to know where the telephones are so that help can be summoned swiftly and to know the whereabouts of the first aid box so that rapid assistance can be given to an injured person. The main taps for gas and water and the switch for electricity should also be located so that these services can be turned off in an emergency.

The person in charge of a class should of course point out where the safety equipment is located and also draw attention to any specific hazards to be found in a particular experiment.

Safety notices

Laboratory workers must also know the meaning of safety signs. Some of these are in plain English while others are in the form of pictograms. The signs have been standardized in Britain and Europe in terms of lettering, diagrams and colour so they can be rapidly identified; some examples of these are given in Fig. 1.1.

Personal protection

Goggles or safety spectacles Eyes are especially vulnerable to splashes from reagents and safety spectacles should always be worn when carrying out any procedures where there is a risk.

Gloves Heavy duty gloves must be worn when handling corrosive substances such as strong acids or alkalis. The hazardous nature of these substances is obvious but the dangers inherent in skin contact with other chemicals are not always clear.

Lightweight disposable gloves should therefore be worn during weighing and handling of chemicals to avoid the risk of absorption through the skin.

Protective clothing   Laboratory coats are not status symbols but are meant to protect the wearer from chemical splashes and infectious material. Cotton is a better material for a lab coat than nylon as it has a greater absorptive capacity and is generally more resistant to chemical splashes.

The standard open-neck coat may be adequate for most chemical work but a high necked gown is more suitable for work with animals and potentially dangerous micro-organisms.

Face masks   These are not always necessary but need to be worn when there is a risk of dust from chemicals or an aerosol of micro-organisms.

Dangers to avoid

Poisoning often arises from the accidental transfer of a compound to the mouth and this risk can be greatly reduced by always keeping three simple rules in the laboratory.

1. No smoking
2. No eating and drinking
3. No mouth pipetting

Colour		Meaning	Examples
Background	Lettering or symbol		
Red	White	Fire fighting	Fire Blanket / Fire alarm
Red or White	White or Red	Prohibition	Keep Out / No Smoking
Yellow	Black	Warning	Radiation / biological hazard
Green	White	Emergency	Exit / First Aid
Blue	White	Mandatory	Now wash your hands
White	Black	General	Private / (toilet symbol)

WATER

USE FOR WOOD, PAPER
FABRICS ETC.

DO NOT USE ON ELECTRICAL
OR FLAMMABLE LIQUID FIRES

FOAM

USE FOR FLAMMABLE LIQUIDS,
OILS, FATS, SPIRITS ETC.

DO NOT USE ON
ELECTRICAL FIRES

FIRE BLANKET

USE FOR SMOTHERING

POWDER

USE FOR ALL RISKS·
FLAMMABLE LIQUIDS, & GASES

CO₂

USE FOR ELECTRICAL &
FLAMMABLE LIQUID FIRES

B C F

USE FOR ELECTRICAL &
FLAMMABLE LIQUID FIRES

Chemical hazards

All chemicals should be considered as potentially dangerous and handled accordingly. Contact with skin and clothing should be avoided and even if a chemical is thought to be harmless it should not be tasted or smelt.

Hazard warning symbols, which are black on an orange background, are present on reagent bottles to warn of specific dangers and must be heeded. Solutions of reagents placed out for classwork should also be marked by the technical staff and coloured adhesive labels are available for this purpose.

Corrosive and irritant substances

Corrosive

Harmful

A corrosive substance is one that destroys living tissue and the inherent dangers of strong acids or alkalis coming in contact with the skin are only too obvious.

An irritant on the other hand cause local inflammation but not destruction of the tissue and the dangers in this case are more subtle and not always appreciated. For example, occasional contact with the skin may suggest that the substance has no detectable effect. However, repeated exposure can suddenly give rise to irritation as in the case of some organic solvents.

Toxic compounds

Toxic

Compounds are graded as toxic or highly toxic depending on the dose required to kill 50 per cent of a population of animals (LD_{50}). The inherent dangers of swallowing a toxic compound are obvious but the dangers of absorption through the skin or inhalation are not always appreciated.

Some compounds take a long time before their toxicity becomes evident and this is particularly true for carcinogens and teratogens. Some common biochemical reagents show this long-term toxicity; ninhydrin for example is carcinogenic and thyroxine is teratogenic. If possible a substitute should always be used but if none is available then extra care must be taken when using these substances.

Flammability hazards

Flammable

Oxidizing

Flammable substances are those with a low flash point and *all* naked flames in the laboratory should be extinguished when handling them and not only those in the immediate vicinity of the substance. Sparks from electrical equipment are less obvious than a Bunsen burner but can be just as dangerous. For this reason, organic solvents must not be stored in the refrigerator.

Oxidizing substances may not be flammable themselves but may cause a fire when brought into contact with combustible material.

The best precaution if such compounds need to be used, is to have only the minimum amount required on the bench and to keep the main bulk in steel cabinets well away from the work area.

Explosives

Explosive

Explosives as such are not handled in the normal biochemical laboratory but some general laboratory reagents such as picric acid are explosive and must be handled with extreme caution.

As with flammable compounds, only small quantities of the compound should be used in the work area and preferably behind a protective screen. Explosions can also arise from the mixture of two compounds which in themselves are harmless and an awareness of this is necessary to avoid a laboratory disaster.

Physical hazards

Fire

Action to take   If you discover a fire then act quickly and follow these simple rules:

1. Don't panic.
2. Raise the alarm.
3. Evacuate laboratory.
4. Turn off gas and electricity.
5. Attack fire with extinguisher.
6. *If in doubt get out!*

Only attempt to tackle the blaze if it is a small one and you know what you are doing. It is vital to recognize the type of fire and to use the correct type of extinguisher. Indeed selection of the wrong extinguisher may actually make matters worse. For example, water-based extinguishers should not be used on solvent fires to avoid spreading the blaze and should never be used if electrical equipment is involved because of the risk of electrocution.

The major cause of death in fires is the inhalation of carbon monoxide and other toxic fumes in the smoke rather than burns, so if there is any difficulty at all in breathing don't try to be a hero, get out!

Types of fires   There are different types of fires and the European classification of these is given below.

A. Solid materials, usually organic
B. Liquids and liquefiable solids
C. Gases
D. Metals

Some safety publications talk about 'electrical fires' but these do not exist as such although all four types of fire may involve electrical equipment.

Gas cylinders   A gas cylinder in a fire is a potential bomb and the best course of action is to evacuate the building immediately and warn the fire brigade. Such fires should always be left to the experts to tackle.

Extinguishers   If the fire is small then it should be tackled with a fire extinguisher or blanket provided there is a safe exit. There are several types of extinguishers available in a laboratory; they can be identified by their colour and the best one for tackling the blaze is selected according to the nature of the fire (Fig. 1.2).

In the case of a fire involving oils or fats then smothering the blaze with a fibre glass blanket is often more effective than using an extinguisher.

Pressure

Low or high pressure can be hazardous although the former is easier to guard against. Vacuum lines, for example, should be run in a fume chamber behind a protective screen while Dewar flasks and vacuum desiccators need to be wrapped in a net in case of an implosion.

High pressure, especially in the form of a gas cylinder, is more of a potential hazard because of the added risk of fire or an explosive reaction.

Use of gas cylinders   Gas cylinders are used with complete safety in many laboratories but they can be dangerous objects. Accidents can happen but these nearly always occur through ignorance and incorrect use. For this reason if you need to use a gas cylinder always get instructions before touching it and if in doubt ask.

Storage of cylinders   Cylinders should be firmly clamped to the bench in an upright position. This makes sure they do not fall on anyone or become a lethal torpedo in the event of an accident.

Identification of cylinders   Unfortunately there is no international colour code of gases so it is vital to check whether the cylinder about to be used is imported and to identify clearly the gas from the lettering on the cylinder and not rely on the colour.

Fuel gases   It is vital not to mix even small amounts of a fuel gas with an oxidizing gas and so cylinders of fuel gases such as hydrogen are fitted with a left-handed thread whereas other cylinders have right-handed threads for attaching the valve and gauge.

Oil and grease   Oil and grease can explode at a high temperature in the presence of an oxidizing gas and so these materials are never used on gas cylinders.

Ionizing radiation

Types of radiation   Radionuclides can give rise to α, β and γ-rays as well as neutrons while some items of high voltage electrical equipment can produce X-rays.

Alpha particles (α)   These are doubly charged helium nuclei ($_2^4\text{He}^{2+}$) and are only emitted by radioactive elements with a very high atomic number. α-Emitters do not represent an external radiation hazard since the α-particles are stopped by a layer of skin but they are extremely dangerous if ingested. The tissue immediately in the vicinity of the compound becomes intensely ionized and this leads to cellular injury and the development of cancer. Fortunately α-emitters are only rarely used in biochemical work.

Beta radiation (β)   This consists of high speed electrons (β^-) with a range of kinetic energies. Low energy particles are absorbed by the outer layers of the skin but those with a high kinetic energy can penetrate up to 3 mm and cause unpleasant skin burns. Many of the isotopes commonly used in biochemistry are β-emitters (^3H, ^{14}C, ^{32}P).

Gamma radiation (γ)   This is electromagnetic radiation of short wavelength and has no mass or charge. It is therefore extremely penetrating and creates areas of ionization following collision with atoms. Some of the isotopes used in biochemical work are γ-emitters (^{131}I, ^{54}Fe).

Neutrons   Neutrons carry no charge and so are highly penetrating and thereby dangerous. However, they are rarely met in the biochemical laboratory.

Radioactive

Warning signs   The symbol for radiation is black on a yellow background (see Fig. 1.1) and must be clearly displayed on the door of a laboratory where radiation could be a hazard whatever the source.

This hazard warning label which is black on an orange background is found on reagent bottles that contain radioactive substances.

Exposure   All radiation must be treated as dangerous and exposure kept to an absolute minimum and always below the maximum permitted dose. It is as well to remember that there is no such thing as a 'safe' dose of radiation.

Personal monitoring   Anyone working in a laboratory with a radiation hazard is required to wear a film badge continuously. This is worn on the trunk for whole body radiation or on the wrist to monitor exposure to the hands. If high energy β- and γ-emitters are used then a pocket dosemeter or alarm is useful as well as the film badge.

Handling   The precautions routinely used when working with toxic materials should be adopted with the added safeguard of shielding with perspex or lead as appropriate.

On completion of the experiment any radioactive waste must not be put down the sink but disposed of according to the instruction given by the demonstrator.

Non-ionizing radiation

Direct exposure to radiation can be avoided with precautions but reflections can be quite dangerous especially if the radiation is invisible.

Visible radiation   Lasers are the biggest hazard from visible radiation and protective goggles must be worn. Lasers are extremely dangerous and can cause blindness from less than one second's exposure.

Ultraviolet   The ultraviolet region of the electromagnetic spectrum that is hazardous to living organisms is from 200 nm to 315 nm with a maximum at 270 nm. These wavelengths are readily absorbed by the nucleic acids of the chromosomes causing long-term damage and genetic mutations. The immediate acute danger is to the eyes where UV can cause conjunctivitis and to the skin where it can give rise to sunburn. These effects can be prevented by wearing glass goggles and using a barrier cream.

Other radiation   The dangers from other forms of radiation are not so well understood but *infrared* and *microwave* radiation are said to cause cataract in the eyes. High intensity *ultrasonic beams* may also cause damage to living tissue but the exact mechanism of this is not yet clear.

In view of the uncertainties, it is best to treat all forms of radiation with care and to take steps to provide adequate protection for the eyes and skin.

Biological hazards

Micro-organisms

Infection can occur from a cut or by ingestion but this can be prevented by taking the precautions already discussed. Less obvious is the risk of infection by inhaling the particles from an aerosol and this is probably the biggest danger when handling bacteria or viruses.

Aerosols are formed whenever a fluid surface is broken, as for example with careless pipetting. Aerosols containing micro-organisms are highly dangerous if they have a diameter of 1–5 μm as this size readily penetrates the lung.

Centrifugation   Aerosols are easily formed by centrifuging unstoppered tubes but this can be prevented by making sure that all tubes are firmly capped. The centrifuge buckets should be balanced with tubes containing disinfectant and not water which might otherwise become infected.

Personal protection   The laboratory coat should be of the surgical type with a high neck and buttoned sleeves. Goggles, gloves and a face mask should also be worn to give maximum protection.

Safety cabinets   Micro-organisms should be handled in a safety cabinet or hood and not in the open laboratory. In these cabinets, air flows vertically away from the user then passes through a filter which must be changed frequently and sterilized.

Flame loops   Wire loops used to plate out cultures can give rise to aerosols if handled carelessly. They should never be overloaded since they can spatter and spread infected material when placed in the flame of a Bunsen.

Animals and body fluids

Infection can also occur when working with tissues, body fluids and whole animals and the same precautions should be used when handling these materials as with the direct manipulation of micro-organisms.

Animals There are a large number of diseases that affect animals and can be transmitted to man so even apparently healthy animals need to be handled with care.

Animals can also cause allergies in some individuals but there is little that one can do about this except to avoid direct exposure as far as possible.

Body fluids Blood and plasma, which are frequently used in biochemical experiments, also need to be treated with caution as they are a potential source of viral infections. The acquired immune deficiency syndrome (AIDS) is probably the most publicized and horrific of these infections but it is still thankfully rare. A much greater risk to the laboratory worker is from serum hepatitis. Infection is effective primarily through the blood but other body fluids can transmit the virus. The mortality rate for one particularly virulent form (Australia antigen) can be as high as 30 per cent of those infected and there is no known cure for the disease. Furthermore, the virus can be carried by apparently healthy individuals so all apparatus in contact with human blood or urine should be sterilized immediately after use. The carrier rate may be as low as 0.1 per cent (British population) or as high as 20 per cent (some African populations). All human fluids should therefore be treated with the greatest care and strict precautions taken.

Genetic engineering

Work with bacteria has always involved the risk of mutations occurring, either spontaneously or following exposure to radiation or chemical agents. Most mutations are harmless but some may be dangerous especially those that are pathogenic or resistant to antibiotics. However, recent advances in molecular biology have raised the possibility of new hazards occurring, particularly those involving the manipulation of genetic information between bacteria.

Recombinant DNA The experiments causing most concern are those that involve recombinant DNA in which a fragment of nucleic acid is removed from a bacterium and incorporated into the nucleic acid of a plasmid or bacteriophage. These recombinant molecules are then cloned in bacteria which are usually *Escherichia coli*.

The biological properties of these new hybrid molecules are unknown and this coupled with the possible transfer to man has given rise to some concern. Working parties were set up by several countries to examine the possible hazards of such experiments and they all concluded that two particular types of experiment could be potentially hazardous.

New pathogenic bacteria *Escherichia coli* are normally present in the human intestine and accidental infection with bacteria containing hybrid DNA could lead to the transfer of genetic information to the gut *E. coli* or to other bacteria pathogenic to man. This would be particularly serious in the case of the bacterial genes for the production of a toxin or antibiotic resistance.

Cancer   The other hazard considered a risk was the possibility of cancer when manipulating oncogenes or hybrid molecules containing animal DNA whose properties would be completely unknown.

So far there is no evidence that these fears are real but work with recombinant DNA should only be carried out after official approval and under very strict safety precautions. At the moment such experiments are unlikely to be carried out in teaching laboratories but students should nevertheless be aware of such dangers.

Spillage and waste disposal

Spillages

Chemicals   Reagent bottles should be carried firmly and well supported while large containers must be transported in a wire basket. The work area needs to be kept as clear as possible by keeping any bottles and flasks away from the edge of the bench and placing them on a shelf when not required. Don't leave pipettes in flasks or bottles which are then very susceptible to a knock.

Spilt chemicals should be mopped up with an absorbent material and the area thoroughly washed with water. If corrosive substances are spilt then they must be neutralized before being cleaned up.

Some chemicals need special treatment and the wall chart published by BDH 'How to Deal with Spillages of Hazardous Chemicals', is a very good source of reference.

Radioisotopes   If the reagent is radioactive then warn any workers in the immediate vicinity and get expert help in dealing with it.

Infected material   Thoroughly wash the area with a solution of disinfectant and report the incident.

Waste disposal

Chemical waste   Small quantities of many materials may be washed down the sink with plenty of water but some reagents must not be disposed of in this way. Chemicals such as acids and alkalis need to be neutralized first before washing down the sink while others can only be dealt with by incineration. It is important that organic solvents are not poured down the sink but stored in metal drums. For safety reasons, the chlorinated solvents must be stored by themselves and not mixed with other organic solvents.

Radioactive waste   The method of disposal depends on the isotope and may be via the sink, incineration or storage for later removal. Never flush radioactive material down the sink unless you are told it is safe to do so and always discuss the method of disposal with the radiation officer.

Infected material No infected material should ever leave the laboratory. Contaminated glassware must be placed immediately after use in a solution of a disinfectant such as sodium hypochlorite containing a dye to show when it is no longer effective. Other material should be sterilized in an autoclave kept specifically for that purpose and should then be disposed of.

Finally, if you are unsure of how to deal with waste material, get expert help and advice from the technical staff.

First aid

First aid as the name implies is the help given immediately to an injured person. It is not a substitute for medical or hospital attention and a doctor or ambulance should be summoned while treatment is being given unless the injury is very minor. It is best to call someone qualified in first aid but this is not always possible and the casualty may require immediate assistance which only you are in a position to give.

The following short notes are not a substitute for a proper course in first aid but should be of some help if you find yourself dealing with the subject of a laboratory accident.

Immediate assistance

Artificial respiration The brain is unable to function for more than a few minutes without oxygen and brain damage or death quickly follows. If the casualty has stopped breathing then artificial respiration must be started at once. It should really be applied by someone trained in the technique but don't wait until help arrives, do something immediately.

1. Lay casualty on back.
2. Clear any obstruction of the mouth.
3. Place a coat under the shoulders so the head is tilted back.
4. Pinch the nostrils and apply mouth to mouth resuscitation 10 times a minute until breathing starts.

Unconciousness If breathing has stopped then apply artificial respiration immediately. If the victim is breathing then lay him face down with his head on one side and the arm and leg of that side in a bent position. This posture makes breathing easier for the patient and provides a better circulation of blood to all parts of the body.

Shock If conscious, keep the casualty lying down with his feet up and cover him with a light blanket to keep warm. Do not leave him alone but stay to reassure him that help is on the way. Do not give any drinks.

Burns

Severe burns If the casualty is on fire act immediately to douse the flames by wrapping him in a fire blanket; then cool the affected area of the body under running water. Remove any contaminated clothing and wrap up the casualty to minimize shock.

Minor burns and scalds Do not touch the burn but wash the affected area under running cold water for at least 15 minutes, then cover with a dry sterilized dressing.

Alkali burns Flush the burn with water for about 15 minutes and apply a dry dressing.

Acid burns Treat the same as alkali burns but bathe the area gently with a solution of sodium bicarbonate before applying the dressing.

Electric burns Switch off the current or if this is not possible free the person using anything that is non-conducting. Check breathing and treat the burns as described.

Some don'ts
Do not remove clothing stuck to a burn.
Do not burst any blisters.
Do not apply any cream or ointment.
Do not apply cotton wool or similar material.
Do not use an adhesive dressing.

Eye injuries

Foreign body A piece of dirt can often be removed by getting the person to blink rapidly. In most cases the movement of the eyelid coupled with an increased flow of tears is usually sufficient to remove the object. If this is unsuccessful, the body can sometimes be removed with the corner of a clean handkerchief. However, if these methods are unsuccessful do not persist but cover the eye with a pad and bandage flat to keep the eye shut until medical assistance is obtained.

Chemicals in the eye Flush the eye with cold water for 15 minutes. Lightly cover the eye with a sterile eye pad and take the casualty to hospital.

Bleeding

Severe bleeding Make the victim lie down and apply direct pressure to the wound with a sterilized pad and a firm bandage. If an artery is cut get someone trained in first aid to apply a tourniquet but act quickly. If an arm or leg is badly cut but not broken, raise the limb above the rest of the body.

Minor cuts   Thoroughly wash the cut with soap and water or antiseptic, remove any foreign bodies, then cover with a sterilized dressing.

Toxic materials

Gassing   Get the casualty out into the fresh air, keep warm and apply artificial respiration if his breathing has stopped.

Poisoning   All poisons should be diluted with water but do not give an emetic if the poison is a corrosive liquid or organic solvent. Keep the patient warm and stay with him until an ambulance arrives. Try to find out what poison was ingested by asking the victim and by examining the area of his work. Such information could be of vital importance and affect the treatment given at the hospital.

Further reading

Bretherick, L., *Handbook of Reactive Chemical Hazards*, 2nd edn. Butterworths, London, 1979.

Bretherick, L., *Hazards in the Chemical Laboratory*, 3rd edn. The Royal Society of Chemistry, London, 1981.

Department of Education and Science, *Safety in Science Laboratories*, 3rd edn. HMSO, London, 1978.

Freeman, N. T. and Whitehead, J., *Introduction to Safety in the Chemical Laboratory*. Academic Press, New York, 1982.

Fuscaldo, A. A., Erlick, B. J. and Hindman, B., *Laboratory Safety, Theory and Practice*. Academic Press, New York, 1980.

Hartree, E. and Booth, V., *Safety in Biological Laboratories*. The Biochemical Society, London, 1977.

Hawkins, M. D., *Technician Safety and Laboratory Practice*. Cassell, London, 1980.

2. Accuracy

Accuracy may be defined as the degree of conformity to the truth. Therefore, the biochemist like all scientists seeks to obtain accurate measurements in the laboratory and also to present an accurate account of the experiment. These aspects of practical work are dealt with at this stage, since only when these requirements are fulfilled will the practical work carried out be of value to the student and to those who will read the report. However, before dealing with these points, it is as well to remind ourselves of the units and quantities used in biochemical experiments and to consider some simple calculations.

SI units

The units used in this book are SI units (*Système International d'Unités*) based on the metric system. These were approved in 1960 by The General Conference of Weights and Measures and are being adopted by scientific laboratories throughout the world. They are a *coherent* system of units so that if two unit quantities are multiplied or divided, then the answer is the unit of the resultant quantity. In this way the number of multiples and submultiples of units now in use will be reduced.

Basic units

There are seven basic units on which all others are based and these are set out in Table 2.1.

Table 2.1 SI units

Physical quantity	Name	Symbol
Length	Metre	m
Mass	Kilogram	kg
Time	Second	s
Amount of substance	Mole	mol
Thermodynamic temperature	Kelvin	K
Electric current	Ampere	A
Luminous intensity	Candela	cd

Derived units

In addition to those above, there are also a number of derived SI units obtained by appropriate combination of these basic units. For convenience, these derived units are

Table 2.2 The special names and symbols of some derived SI units

Physical quantity	Name	Symbol	Units
Frequency	Hertz	Hz	s^{-1}
Force	Newton	N	$kg\,m\,s^{-2}$
Pressure	Pascal	Pa	$N\,m^{-2}$
Energy or work	Joule	J	$N\,m$
Power	Watt	W	$J\,s^{-1}$
Electric charge	Coulomb	C	$A\,s$
Electric capacitance	Farad	F	$A\,s\,V^{-1}$
Potential difference	Volt	V	$W\,A^{-1}$
Resistance	Ohm	Ω	$V\,A^{-1}$
Conductance	Siemens	S	Ω
Radioactivity	Becquerel	Bq	S^{-1}
Absorbed dose of radiation	Gray	Gy	$J\,kg^{-1}$
Customary temperature	Degree Celsius	°C	$°C = K - 273.15$

None of these units takes the plural form so that 5 volts is written 5 V not 5 Vs and 2 metres is written 2 m not 2 ms.

given special names and those which are likely to be met in biochemical work are listed in Table 2.2.

Prefixes

Sometimes units may be too large or too small and, in this case, in order to avoid writing too many zeros, a prefix is placed before the symbol of the unit. The recommended multiples or fractions of a unit change by 1000 each time (Table 2.3). Thus 0.000 015 mol is written 15 μmol and 13 400 m is written 13.4 km.

The combination of prefix and symbol is regarded as a new symbol so that mm^3 means $(10^{-3}\,m)^3$ not $10^{-3}\,m^3$. There is therefore no space, point or full stop between the prefix and the symbol. However, a space is left between symbols in derived units and for the sake of clarity a point above the line is often used. For example:

$$ms = millisecond\ (i.e.)\ 10^{-3}\,s,$$

whereas

$$m \cdot s = metre \times second\ (i.e.)\ m \times s$$

Table 2.3 Common prefixes for the SI units likely to be used in biochemical work

Multiples			Fractions		
Factor	Prefix	Symbol	Factor	Prefix	Symbol
10^6	mega	M	10^{-3}	milli	m
10^3	kilo	k	10^{-6}	micro	μ
			10^{-9}	nano	n
			10^{-12}	pico	p

Table 2.4   Some universal constants

Constant	Symbol	Value	Units
Avogadro's number	N_A	6.023×10^{23}	$kmol^{-1}$
Gas constant	R	8.314×10^3	$J\,kmol^{-1}\,K$
Boltzmann constant	k	1.381×10^{-23}	$J\,K^{-1}$
Charge on the electron	e	1.602×10^{-19}	C
Faraday	F	9.649×10^{-7}	$C\,kmol^{-1}$
Planck's constant	h	6.626×10^{-34}	$J\,s^{-1}$
Gravitational constant	g	9.81	$m\,s^{-2}$
Velocity of light in a vacuum	c	2.998×10^8	$m\,s^{-1}$

Physical constants

Some of the universal constants which the biochemist is likely to encounter are given above together with their values in SI units (Table 2.4).

Units used in conjunction with SI

There are some units used in biochemical work which are unlikely to be replaced altogether because of their convenience and it is probable that they will continue to be used in conjunction with SI units for some time.

Litre (l)

The coherent SI unit for volume is, of course, the cubic metre (m^3) but this is rather large and the litre is still accepted as the unit of volume in biochemical work.

The litre is exactly equal to one cubic decimetre (1 decimetre = 10^{-1} m = dm) so that:

$$1000 \text{ litres} = 1 \text{ cubic metre} = m^3$$
$$1 \text{ litre (l)} = 1 \text{ dm}^3 \qquad = 10^{-3}\,m^3$$
$$1 \text{ millilitre (ml)} = 1 \text{ cm}^3 \qquad = 10^{-6}\,m^3$$
$$1 \text{ microlitre (}\mu l\text{)} = 1 \text{ mm}^3 \qquad = 10^{-9}\,m^3$$

The terms millilitre and microlitre will be abandoned in time but probably not until after the useful life of this text.

Gram (g)

The gram will continue to be used as an elementary unit and in association with prefixes (μg, mg) until a new name is adopted for the basic unit of mass now known as a kilogram.

Minute

The basic SI unit for time is the second but the common units of time (e.g., minute, hour, year) can still be used when convenient.

Molarity and moles

In the experience of the author, difficulty is often encountered by students over moles and molarity and calculations involving conversions from molarities to millimoles or micromoles in a given volume. This section should therefore be read and understood before attempting any of the experiments involving calculations.

Mole (mol)

The basic SI unit of quantity is the mole which gives the *amount* of a substance present in, say, a flask or test tube irrespective of the volume present. It is defined as the molecular weight of a compound in grams

$$1 \text{ mole} = \text{molecular weight in grams} = 6 \times 10^{23} \text{ molecules}$$
$$\text{(Avogadro's number).}$$

The term mole is also applied to other particles of defined composition such as atoms, ions, or free radicals, as well as molecules:

1 mole of glucose (mol. wt 180) is 180 g,

1 mole of albumin (mol. wt 68 000) is 68 000 g or 68 kg.

Molarity (mol/litre)

The amount of a substance present in unit volume of solution gives the *concentration* of that substance and, in biochemical work, the unit amount is the mole and the unit volume the litre. A molar solution of a compound is therefore defined as 1 mole of that compound per litre.

$$1 \text{ mol/litre} = \text{mol. wt in grams per litre of solution}$$

Molarities have been written using M as the symbol (0.15 M NaCl) but the SI recommendations are that this be replaced by mol/litre (NaCl 0.15 mol/litre).

Mass concentration

Sometimes, measurements are made on substances which do not have a defined composition, such as the concentration of protein or nucleic acid present in an extract. In these cases the concentration is expressed in terms of weight per unit volume rather than moles. It is also used when the molecular weight of the biologically active compound in a mixture is uncertain, as in the case of vitamin B_{12} and serum immunoglobulins. The unit of volume is still the litre, so all concentrations should be expressed with the litre (g/litre, mg/litre, µg/litre, etc.) and not 100 ml as the base. The term 'per cent' is still used but should be discontinued, unless clearly defined, because of its ambiguity. For example, a 2 per cent solution of acetic acid could mean:

2 g of acetic acid per 100 g of water (w/w),

2 g of acetic acid per 100 ml of water (w/v),

2 ml of acetic acid per 100 ml of water (v/v).

Relationship of molarity and moles

In many biochemical reactions the number of moles of a substance in the test tube needs to be known, and this can be readily calculated from the molarity of the solution and the volume present. To do this the number of moles present in 1 ml is first calculated, then multiplied by the volume of the solution present. The following relationship is worth remembering and is obtained by decreasing both the amount and volume by a factor of 10^3 each time.

$$A \text{ molar solution} = 1 \text{ mol/litre}$$
$$= 1 \text{ mmol/ml}$$
$$= 1 \text{ μmol/μl.}$$
$$\text{Similarly, a millimolar solution} = 1 \text{ mmol/litre}$$
$$= 1 \text{ μmol/ml.}$$

Simple biochemical calculations

It is important to understand the difference between moles and molarity and to be able to express the answers to laboratory calculations in the correct units. The concepts discussed above are really quite simple yet students can and do get muddled and this can persist even into the final year when units should be a matter of instinct rather than thought.

Remember:

moles = an amount

molarity = a concentration, i.e., an amount in a known volume (litre).

Don't confuse these terms!

The Michaelis constant of an enzyme is a concentration and so:

$$K_m = 150 \text{ μmol/litre} \quad NOT \quad K_m = 150 \text{ μmol}$$

The activity of an enzyme is expressed as μmol of substrate changed per minute and so:

$$v = 13 \text{ μmol/min} \quad NOT \quad v = 13 \text{ μmol/litre/min}$$

Check that you have grasped these ideas by trying the following calculations, the answers to which are given in the Appendix.

1. How many grams of glucose are needed to make 100 ml of a 0.6 mol/litre solution? (Glucose mol. wt = 180)
2. How many millimoles or micromoles per millilitre are present in the following solutions: (a) 0.15 mol/litre NaCl; (b) 12 mmol/litre fructose; (c) 200 μmol/litre ATP; (d) 30 g/100 ml urea (mol. wt 60)?
3. How many grams of glycine are there present in 10 ml of a 20 mmol/litre solution? (Glycine mol. wt = 75)

4. If the fasting blood glucose in man is 5 mmol/litre, what is the concentration expressed as mg per 100 ml of blood?
5. A sample of plasma (0.2 ml) was mixed with 0.8 ml of distilled water and 2 ml of biuret reagent. The extinction of the mixture after 10 min was the same as that obtained from mixing 1 ml of albumin (1.3 mg/ml) and 2 ml of the biuret reagent. Calculate the concentration of protein in the plasma.
6. The enzyme activity of a tissue extract was determined by following the increase in extinction of the product as it is formed. After 10 min, the extinction measured was the same as that of a 30 μmol/litre standard solution of the product. If the volume of tissue extract assayed was 0.2 ml in a total volume of 3 ml, calculate the enzyme activity of the tissue extract in terms of nanomoles of product formed per minute.

Accurate measurement

The sources of error

No measurement taken in the laboratory is exact; all measurements are of limited accuracy and are liable to errors so the potential source of these errors should be appreciated.

An error is the amount of deviation from the correct or true value and not a mistake on the part of the worker. Errors arise from statistical variation and have to be lived with although they can be minimized by careful working. Mistakes on the other hand are instances of human incompetence which can and must be eliminated.

Human mistakes   These can arise from a badly designed experiment where insufficient control is exercised. For example, in many biological experiments the temperature and illumination of the environment may have profound effects on the system under investigation and should therefore be carefully controlled. Experiments should be planned in such a way that only one variable is introduced at a time and all other factors that may affect the experiment are kept constant.

A common source of human error arises from the careless reading of a scale or meniscus when the problem of parallax is not appreciated. For this reason, many instrument scales incorporate a mirror behind the pointer so the true reading is obtained when the pointer and its reflection are superimposed. Some manufacturers overcome this problem by giving a digital readout on their instruments rather than a deflection on a scale. Probably one of the greatest sources of human error has been the careless reading of a meniscus on a pipette (Fig. 2.1) but this has now been eliminated by the introduction of automatic or digital pipettes, e.g., Gilson, Eppendorf, etc.

Limitations of apparatus   The limits set by the accuracy of a particular piece of equipment are usually known and should always be allowed for. For example, the error on a graduated 10 ml pipette may be 0.2 per cent, in which case the pipette can deliver a large volume, e.g., 9.2 ml, quite accurately, but the error on trying to deliver 0.1 ml would be as high as 20 per cent. These errors are also known and can be taken into account.

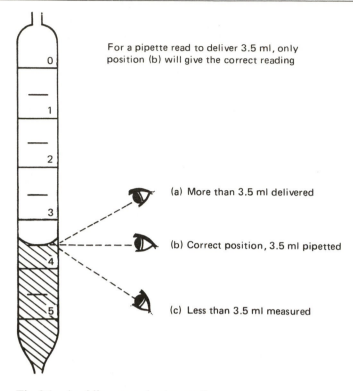

For a pipette read to deliver 3.5 ml, only
position (b) will give the correct reading

(a) More than 3.5 ml delivered

(b) Correct position, 3.5 ml pipetted

(c) Less than 3.5 ml measured

Fig. 2.1   Avoiding error due to parallax

Standards and blanks   To obtain as accurate a value as possible from an estimation, errors must be reduced to a minimum and this can be done by careful working and the use of standard solutions. Standard solutions of the substance to be estimated should be included with any test, even when a calibrated instrument and standard reagents are used. This provides a useful check on the accuracy of a method since the measured figures should fall within the acceptable limits of the true values. Ideally the standard solution should be treated in an identical manner to the fluid under investigation. A standard curve can then be constructed showing the variation of the quantity measured with concentration. Values obtained for the test solution should fall within the range of the standard curve and the value of the test can then be read. Usually, only one standard is included for volumetric estimations or when a standard curve has previously been constructed.

Blank solutions should be included in any measurements. The same volume of distilled water replaces the substance to be estimated and the blank is then treated in exactly the same way as the test and standard. Any value obtained for the blank is, of course, subtracted from the value of the test and standard in the final calculation, since the blank value is due to the reagents used and not the substance under investigation. The practical use of blanks and standards is well illustrated in the numerous colorimetric estimations in this volume. Several blanks or controls need to be used when working with enzymes, but these are considered separately in the section on enzymes.

Random errors Finally there are random errors which are individually unpredictable. These are seen when one person carries out a number of determinations under identical conditions and obtains a slightly different result each time. This random error can be considerably reduced by taking a large number of measurements and calculating the average value (Table 2.5). For many purposes, duplicate estimations are sufficient provided there is good agreement between them, and this is usually the situation for readings obtained in, say, the construction of a calibration curve. The degree of agreement between replicate experiments is termed the precision. Precision does not mean accuracy, since measurements may be highly precise but inaccurate due to a faulty instrument or technique.

Table 2.5   Determination of serum chloride

| Estimation number | Chloride (mmol/l) | |
	Experimental value	Average value
1	102	102
2	104	103
3	106	104
4	104	104
5	103	104
6	105	104

In many cases, however, the precision is not as good as that shown in the example (Table 2.5) and there is a much greater spread of results. It is, therefore, useful to be able to give some measure of the spread of readings obtained and, in order to do this, some elementary concepts in statistics are now introduced.

A number of equations are given which are obtained from the theory of statistics and these equations are used as tools with no attempt made to derive them.

The normal distribution curve

If a very large number of readings are taken of some quantity (x) then a curve can be constructed which shows the number of readings related to the value of x (Fig. 2.2). This normal, or Gaussian, distribution of results has a number of characteristics:

1. The *mean* value is $\bar{x}$ and the curve is symmetrical and has a maximum at this value.
2. The point of inflexion occurs at $\bar{x} + \sigma$ and $\bar{x} - \sigma$ so that 68 per cent of all values lie in the cross-hatched range $\bar{x} \pm \sigma$.
3. The curve is such that 95 per cent of all values will be in the shaded range $\bar{x} \pm 2\sigma$ and 99 per cent of all values in the range $\bar{x} \pm 3\sigma$.

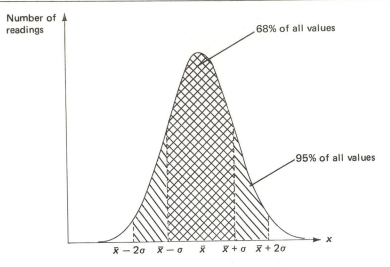

Fig. 2.2   A normal or Gaussian distribution of readings

Standard deviation (SD)   The value of σ is the standard deviation and is a measure of the spread of expected results: σ can be calculated from the individual results $(x_1, x_2, x_3, \ldots, x_n)$, the number of readings taken (n) and the mean $(\bar{x})$.
Now

$$\bar{x} = (x_1 + x_2 + x_3 + \ldots + x_n)/n = \Sigma x_n/n$$

If the deviation of each sample from the mean is represented by d, then:

$$d_1 = x_1 - \bar{x}$$
$$d_2 = x_2 - \bar{x}$$
$$d_3 = x_3 - \bar{x}$$
$$\vdots \qquad \vdots \qquad \vdots$$
$$d_n = x_n - \bar{x}$$

The sum of the squared deviations is known as the *deviance*, and, when this is divided by the number of samples, then the *variance* (σ^2) is obtained.

$$\sigma^2 = (d_1^2 + d_2^2 + \ldots + d_n^2)/n$$

The square root of the variance than gives the *standard deviation* (σ).

Now the mean $(\bar{x})$ and the standard deviation (σ) cannot be known precisely, unless an infinite number of measurements are made. In practice, only a limited number of measurements are possible and a close estimate of the variance of the population can be calculated from a finite number of readings by dividing the deviance by the number of degrees of freedom $(n-1)$, rather than by the size of the sample (n).

It is tedious to calculate the deviation of each individual reading from the mean and the variance, and hence the standard deviation can be more readily determined from the sum of the values of x (Σx) and the sum of the values of x^2 (Σx^2) using the following working formula:

$$\sigma^2 = \frac{\Sigma x^2 - (\Sigma x)^2/n}{n-1}$$

Standard error of the mean (SEM)   Usually, however, a mean value is obtained from the individual readings and it is more important to know how far this value lies from the unknown mean of the whole population than to know the spread of results. The standard error of the mean (σ_m) gives an estimate of the probable error in determining the mean of the population from a finite number of samples.

$$\sigma_m = \sigma/\sqrt{n}$$

From this it can be seen that the larger the number of samples then the smaller will be the SEM and the closer to the 'true' mean of an infinite number of readings.

Coefficient of variation (CV)   Another useful measure of precision is the coefficient of variation. This is the standard deviation (σ) expressed as a percentage of the mean value ($\bar{x}$).

$$CV = \sigma/\bar{x} \times 100 \text{ per cent}$$

The smaller the value of CV, the greater the precision of the measurement.

Biological variation

A physical quantity, such as the density or viscosity of a pure liquid, can be measured in the laboratory and the value obtained compared with the correct figure. Some random variation will be observed, but with care this should be very small, so that only a few measurements need to be taken and the mean calculated. However, this is not the case for many measurements made in biology, where there is often no single 'true' value but a range of so-called 'normal' values. For many measurements a symmetrical 'normal' type of distribution is followed and simple statistical methods can be applied. In this case, the normal range is usually taken to start at ($\bar{x} - 2\sigma$) and extend to ($\bar{x} + 2\sigma$), which would include 95 per cent of all values (Fig. 2.2). Sometimes a 'skewed' distribution is seen which requires more complex mathematical treatment.

The Student t test   In some biochemical experiments it is important to know whether an experiment has caused a significant change in a measured quantity or whether the value obtained is due to chance. An English statistician, who signed himself 'Student', devised a simple test for determining the probability whether a sample belongs to a given population or not by taking into account the spread of results often found.

We have seen that, for a very large number of measurements, the standard error of the mean is close to that of the population ($\bar{x} \pm \sigma_m$), but most experiments involve relatively small samples and the mean of the sample (m) is related to the true mean:

$$m = \bar{x} \pm t \cdot \sigma_m$$

or

$$m = \bar{x} \pm t \cdot \sigma/\sqrt{n},$$

where σ is the standard deviation of the sample and n the number of measurements taken.

Statistical tables have been prepared, and the probability of t can be looked up using the number of degrees of freedom ($n-1$). A probability of 0.05 means that there is a 5 per cent chance that the sample is the same as the population. Similarly if the value of t falls under the 0.01 level of probability then this means that there is only a 1 per cent chance that the sample is the same as the population. These two levels are the ones usually adopted as the *confidence limits* in biology. The results are considered to be *significant* at the 0.05 level and *highly significant* at the 0.01 level of probability.

Two samples consisting of a finite number of measurements each can be compared directly and a value of t obtained using the following equation:

$$t = (\bar{x}_1 - \bar{x}_2)\Bigg/\left[\frac{(n_1-1)\sigma_1^2 + (n_2-1)\sigma_2^2}{n_1+n_2-2}\left(\frac{1}{n_1}+\frac{1}{n_2}\right)\right]^{1/2}$$

If the samples contain the same number of measurements ($n_1 = n_2$) then this simplifies to:

$$t = (\bar{x}_1 - \bar{x}_2)/[\sigma_1^2 + \sigma_2^2/n]$$

Significant figures

The final result of any assay should indicate the accuracy of the measurement, so that a serum calcium of 112 mg/litre means that the concentration is less than 113 mg/litre but more than 111 mg/litre while 111.8 mg/litre means that the serum calcium lies between 111.7 and 111.9 mg/litre. The final result of an assay should include all the significant figures, that is all the certain digits and the first doubtful one. As a general rule when rounding off numbers, if the last figure dropped is 5 or more, add 1 to the last figure retained. A serum calcium of 111.8 mg/litre is therefore expressed as 112 mg/litre to three significant figures.

Electronic calculators generate results with a large number of digits and students should resist the temptation to put them all down. Always round the final results off to the number of significant figures but do not round off the experimental results or the intermediate values otherwise considerable errors may arise in the final result.

Writing up the experiment

The obtaining of accurate results is not an end in itself. The object of laboratory work is to communicate results and ideas to others in a form in which they are understood. The writing up of a laboratory exercise is good practice for the more exacting task of producing a scientific paper (Booth, 1975).

The recording of results

Practical book   A large, stiff, covered book is probably most convenient, although loose leaves in a folder are a possible alternative. In the latter case, there is the advantage that chart traces and graphs can be inserted, but there is the disadvantage that sheets of writing or data can be lost. Records do get lost sometimes and so the name and address of the owner should be clearly written on the first page. My own research assistants also record in their books that an appropriate reward will be given for the safe return of their records should they be lost and students may wish to do this, depending on how highly they value their laboratory results.

The course and departments are also indicated and, after this, several pages left for a contents list which should give the date, title and page number of the experiment.

There are various ways of presenting an account of an experiment and the following headings are those found in most biochemical research papers. In some experiments it may be convenient to combine two sections under one heading, such as Methods and results or Results and discussion, but this depends on the particular investigation.

Introduction   All experiments have a title and this should be put at the top of the page, together with the date. The title should be concise and informative so that the object of the experiment is clearly understood.

'State your objectives'. Such words should be in the forefront of all student minds, and it is good practice to give a brief account of what you are seeking to show in the experiment.

Materials or apparatus used   An account should be given of reagents and equipment used, together with any relevant diagrams of specialized apparatus. In the case of chemicals, avoid trade names or trivial names that are not accepted.

Experiment or method   This describes what you actually did in the operational order and should not be just a copy from a practical book or schedule. It should be written in the past tense and not the first person. An experiment in a book may give instructions as follows:

'Add five drops of test solution to 2 ml of Benedict's reagent in a test tube and heat for 5 min in a boiling water bath. Note the colour change and precipitate formed.'

The above instructions must not be copied into the practical book verbatim, but should be presented as follows:

'Five drops of the test solution were added to 2 ml of Benedict's reagent in a test tube. The solution was heated for 5 min in a boiling water bath and the colour change and precipitate noted.'

The description of the experiment must be concise but sufficient information should be given so that others can repeat what you have done.

Results   It is hoped that some results follow the performing of an experiment and these are presented next. The results should state what you saw, not what the book said you should have observed. Full details should be given of your results. It is not sufficient to report that on performing a particular test a yellow precipitate formed. What was the exact colour of the precipitate: was it bright yellow, yellow–orange, etc.? Was the precipitate heavy, light, gelatinous, or granular? When did the precipitate form: immediately, slowly, in the hot, or in the cold? These points may seem obvious, but are not always appreciated. If the student anticipates doing research later, then it is vital to be trained to report what is actually found during an experiment. In research, it is often the unexpected observation that proves the most fruitful, and an accurate record is needed so that work can be repeated.

A report may read as follows: 'A rat was killed and the liver removed.' This is quite inadequate. The strain, sex, age and weight should be given. Was the rat starved or did it have access to food and water? Was it pretreated in any way? How was the rat killed? A knowledge of the above factors may be quite vital in the interpretation of results and must therefore be reported.

Discussion and conclusion   The discussion should not be a repetition of the results, but a logical argument based on them. It is often useful to formulate a question in the introduction to the experiment, then to see how far this has been answered in the discussion.

A conclusion should always be drawn and this should be brief and to the point.

Tables and illustrations

Tables   Experimental results are conveniently summarized in the form of a table or a graph, depending on the nature of the data. Tables should be numbered consecutively and given an informative heading. In some cases it may also be useful to give further details in a legend immediately under the title. The units in which the results are expressed should be given at the top of each column of figures and not repeated on each line of a table. These units should be adjusted so that a convenient number of digits is present in the table. A concentration of 0.007 2 mol/litre is best expressed as 7.2 under the heading 'concentration (mmol/litre)' or as 72 under the heading '$10^4 \times$ concentration (mol/litre)'.

Illustrations   A brief sketch of specialized apparatus is often worth including in the report. Furthermore, an illustration showing chromatography or electrophoresis results, or a flow diagram of a purification procedure, can convey more information than a lengthy verbal description. Generally speaking, data are best presented in the form of a graph as a larger number of observations can be recorded than with tables. It is also easier to assimilate results from a graph than a table. For example, how well the points fit on a smooth curve will give some idea of the random errors of the experiment. Furthermore, a graph will clearly show a discontinuity in the measurements which may not be readily apparent from a table of figures.

Straight-line graphs   If y is related to x in a manner similar to the equation

$$y = mx + c$$

then a plot of y against x will yield a straight line. The slope of the line will be m and the intercept on the y axis will be c (Fig. 2.3).

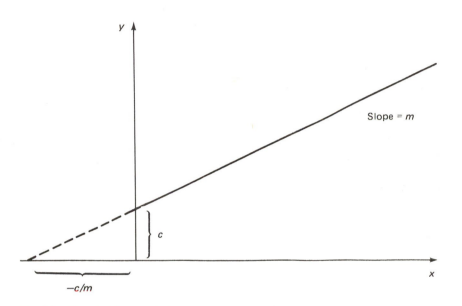

Fig. 2.3   Graph of a straight line ($y = mx + c$)

In many cases the relationship between y and x is not linear, but it is often convenient to manipulate the data in order to obtain a straight-line graph. Some examples of the manipulation of data so as to give a straight-line plot are given in this book: the Beer–Lambert law (Chapter 7) and the section on enzyme kinetics (Chapter 12).

How to present a graph   In many experiments, one quantity such as concentration, pH, or temperature is systematically changed and the effect of this on another quantity is then measured. The known quantity is called the *independent variable* and the unknown or measured quantity the *dependent variable*. When drawing a graph it is customary to plot the independent variable on the horizontal (x) axis (the abscissa) and the dependent variable on the vertical (y) axis (the ordinate). A few hints on how to present a graph are now given.

Identification
1. Give a simple and concise title.
2. Label all axes clearly with the quantities and units.

Scales

3. For the sake of clarity, adjust the scales so that the gradient is about 45°.
4. Use simple numbers (i.e., 15 mmol/litre is preferable to 0.015 mol/litre or 15 000 µmol/litre).

Symbols

5. Use clearly defined symbols (○, ●, □, ■, △, ▲) and not ×, + or a small point ·.
6. Give an indication of the probable error from the size of the symbols.
7. As far as possible have equal spacing between the points.

Curve

8. Join the points together with a smooth continuous curve or a straight line.
9. Draw the curve with instruments and not free hand.
10. Do not run the lines through the symbols.

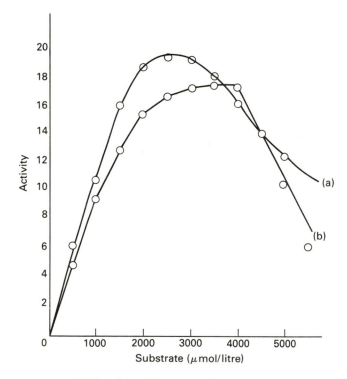

Fig. 2.4   AChE activity. How *not* to plot a graph

Finally, Fig. 2.4 shows how *not* to plot a graph. In the light of the guidance given above and your own common sense, see how many of the 12 mistakes you can spot then check your list with that given in the Appendix.

Experiments

Experiment 2.1   The calibration of laboratory pipettes

PRINCIPLE
The volume delivered by the pipette is determined by weighing the amount delivered and dividing this by the density of water at room temperature (see Vogel A. I., *Textbook of Quantitative Inorganic Analysis*).

MATERIALS	100
1. Balance	20
2. Micropipettes	50
3. Graduated pipette (5 ml)	50

METHOD
(a) Fill the graduated pipette to the 5 ml mark, carefully wipe the end and allow the contents to drain into a weighed bottle. Carefully note the temperature of the liquid and calculate the volume delivered from the weight and density of water at that temperature. Repeat this 10 times and tabulate the results. (b) Set the micropipette to 100 µl, draw distilled water up into the tip and pipette into a small weighed beaker. Calculate the volume actually delivered as given above and repeat with the same tip 10 times.

Accuracy and tolerance limits   Determine the mean volume delivered and calculate the standard deviation of the results. What values would you use as 'tolerance limits' for the pipettes and why?

Low temperature   Many biochemical experiments involve the pipetting of ice-cold solutions, so use your calibrated pipettes to deliver distilled water at 0°C. Calculate the actual volume delivered as before, compare this to the calibrated volume and calculate the error involved in pipetting ice-cold solutions.

Experiment 2.2   The accuracy and precision of the assay for inorganic phosphate

PRINCIPLE
Biochemical experiments are prone to error caused by inaccurate diluting, pipetting, variable times of incubation and the operator's inexperience. The precision is also limited by the quality of the volumetric apparatus and the efficiency of the measuring equipment. These errors accumulate and so the final result always shows some variability and is unlikely to be completely accurate.

In this experiment, the precision and coefficient of variation of the method is determined by each pair of students repeating the assay five times. The effect of human error can also be demonstrated by comparing the results obtained by each pair on the same sample of blood plasma.

MATERIALS   <u>100</u>
1. Materials for the assay of inorganic phosphate   (see Exp. 7.3)
2. Human blood plasma   100 ml
3. Phosphate solution (0.5–0.9 mg P per 100 ml)   100 ml
   (The precise concentration is known only to the person in
   charge and is prepared by the technician by accurately
   diluting the top phosphate standard)

METHOD

Measure the amount of inorganic phosphate in the phosphate solution provided and repeat this five times. Use these results to calculate the accuracy and coefficient of variation of the assay.

Assay the amount of inorganic phosphate in the sample of human blood plasma and prepare a table showing the results obtained by other groups.

Further reading

Booth, V., 'Writing a Scientific Paper', *Biochem. Soc. Trans.*, **3**, 1–26, 1975.

Booth, V., *Communicating in Science: Writing and Speaking*. Cambridge University Press, 1985.

Cornish-Bowden, A., *Basic Mathematics for Biochemists*. Chapman and Hall, London, 1981.

Dawes, E. A., *Quantitative Problems in Biochemistry*, 6th edn. Longman, London, 1980.

Jerrard, H. G. and McNeill, D. B., *A Dictionary of Scientific Units*, 4th edn. Chapman and Hall, London, 1980.

Vogel, A. I., *Textbook of Quantitative Inorganic Analysis*, Longman, London, 1978.

3. pH and buffer solutions

Acids and bases

Definitions

Acids and bases The modern concept of acids and bases developed by Brönsted, Lowry and others defines *acids* as proton donors and *bases* as proton acceptors. Each acid therefore has a *conjugate base*.

$$\text{Acid} \rightleftharpoons \text{Base} + H^+$$

Alkali The term *alkali* is reserved for those compounds that yield hydroxyl ions on dissociation.

$$\text{KOH} \longrightarrow K^+ + OH^-$$

Ampholytes Some ionic species can act as both acids and bases (Table 3.1) and these are known as *ampholytes* or are said to be *amphoteric*.

Table 3.1 Some examples of acids and bases

Acid	Conjugate base
$HCl \rightarrow$	$H^+ + Cl^-$
$CH_3COOH \rightleftharpoons$	$H^+ + CH_3COO^-$
$NH_4^+ \rightleftharpoons$	$H^+ + NH_3$
$H_2CO_3 \rightleftharpoons$	$H^+ + HCO_3^-$
$HCO_3^- \rightleftharpoons$	$H^+ + CO_3^{--}$
$H_2O \rightleftharpoons$	$H^+ + OH^-$
$H_3O^+ \rightleftharpoons$	$H^+ + H_2O$

Although it is convenient to write acid–base equilibria as shown above, the proton does not exist as such, but is usually solvated. For example, in aqueous media the hydrogen ion exists as the hydronium ion (H_3O^+).

$$H^+ + H_2O \rightleftharpoons H_3O^+$$

Strength

Strong acids or bases These compounds are completely ionized in solution, so that the concentration of free H^+ or OH^- is the same as the concentration of the acid or base.

$$\text{Strong acid (nitric acid)}\quad HNO_3 \longrightarrow H^+ + NO_3^-$$

$$\text{Strong base (sodium hydroxide)}NaOH \longrightarrow Na^+ + OH^-$$

Weak acids or bases   These compounds dissociate only to a limited extent and the concentration of free H^+ and OH^- depends on the value of their dissociation constants.

$$\text{Weak acid (formic acid)}\quad HCOOH \rightleftharpoons H^+ + HCOO^-$$

$$\text{Weak base (aniline)}\ C_6H_5NH_2 + H^+ \rightleftharpoons C_6H_5NH_3^+$$

In biological systems, we are dealing with aqueous media and, in this volume, the strength of an acid is taken to refer to water as solvent.

Hydrogen ion concentration and pH

Definition of pH

The hydrogen ion concentration of most solutions is extremely low and, in 1909, Sørenson introduced the term pH as a convenient way of expressing hydrogen ion concentration which avoids the use of cumbersome numbers. pH is strictly defined as the negative logarithm of the hydrogen ion activity, but in practice the hydrogen ion concentration is usually taken and this is virtually the same as the activity except in strongly acid solutions.

The value of using pH can be seen in the case of human blood which has an extremely low hydrogen ion concentration:

$$\text{plasma } H^+ = 0.398 \times 10^{-7}$$

$$\text{plasma } pH = -\log(0.398 \times 10^{-7}) = 7.4$$

Dissociation of water

Derivation of K_w From conductivity measurements, water has been shown to be very weakly ionized and at 25°C the concentration of hydrogen ions is only 10^{-7} mol/litre.

$$H_2O \rightleftharpoons H^+ + OH^-$$

The equilibrium constant for the dissociation of water is given by:

$$K = \frac{[H^+][OH^-]}{[H_2O]}$$

Now the concentration of water to all intents and purposes is constant, so we can write:

$$K_w = [H^+][OH^-]$$

Table 3.2 The ionic product of water and pH of neutrality at various temperatures

°C	K_w	pH of neutrality
0	$0.12 \times 10^{-14} = 10^{-14.94}$	7.97
25	$1.03 \times 10^{-14} = 10^{-14.00}$	7.00
37	$2.51 \times 10^{-14} = 10^{-13.60}$	6.80
40	$2.95 \times 10^{-14} = 10^{-13.53}$	6.77
75	$16.90 \times 10^{-14} = 10^{-12.77}$	6.39
100	$48.00 \times 10^{-14} = 10^{-12.32}$	6.16

The ionic product of water at 25°C is 10^{-14}, so that the pH of pure water at 25°C is 7.

$$[H^+] = [OH^-] = 10^{-7}$$

$$pH = -\log_{10}[H^+] = 7$$

Temperature and K_w At other temperatures, the pH at neutrality is not 7 since K_w varies with temperature (Table 3.2). Even a small change in temperature from 37 to 40°C causes an 8 per cent increase in hydrogen and hydroxyl ions so that a slight rise or fall in temperature may produce a profound biological change in a living system sensitive to hydrogen ion concentration.

Dissociation of acids and bases

Strong acids

These are compounds in which complete dissociation to hydrogen ions and the conjugate base occurs, so that the hydrogen ion concentration is the same as that of the acid. The pH of such solutions can, therefore, be very easily calculated:

(a) 0.01 mol/litre HCl, $pH = -\log_{10}(10^{-2}) = 2$

(b) 0.1 mol/litre HCl, $pH = -\log_{10}(10^{-1}) = 1$

(c) 0.01 mol/litre NaOH, $[H^+] = \dfrac{K_w}{[OH^-]} = \dfrac{10^{-14}}{10^{-2}} = 10^{-12}$

$$pH = -\log_{10}(10^{-12}) = 12$$

Weak acids

The Henderson–Hasselbalch equation   Weak acids are only slightly ionized in solution and a true equilibrium is established between the acid and the conjugate base. If HA represents a weak acid, then:

$$HA \rightleftharpoons H^+ + A^-$$

According to the law of mass action, K_a the acid dissociation constant is defined as:

$$K_a = \frac{[H^+][A^-]}{[HA]}$$

and

$$[H^+] = \frac{K_a[HA]}{[A^-]}$$

Taking negative logarithms,

$$-\log_{10}[H^+] = -\log_{10} K_a + -\log_{10}\frac{[HA]}{[A^-]}$$

or

$$pH = pK_a + \log_{10}\frac{[A^-]}{[HA]}$$

In general terms,

$$pH = pK_a + \log_{10}\frac{[\text{Conjugate base}]}{[\text{Acid}]}$$

The activities of A^- and HA are not always known, so it is convenient to express A^- and HA as concentration terms. Thus:

$$pH = pK_a + \log\frac{C_{A^-}}{C_{HA}} + \log\frac{f_{A^-}}{f_{HA}}$$

where f_{A^-} and f_{HA} are the activity coefficients of A^- and HA respectively. Since $\log(f_{A^-}/f_{HA})$ is constant for a given acid, these activity coefficients can be incorporated into the pK_a term of given an *apparent dissociation constant* pK'_a.

$$pH = pK'_a + \log\frac{C_{A^-}}{C_{HA}}$$

This relationship is known as the Henderson–Hasselbalch equation and is valid over the pH range 4–10 where the hydrogen and hydroxyl ions do not contribute significantly to the total ionic concentration.

pK_a This, as previously defined, is the negative logarithm of the acid dissociation constant of a weak acid. Another way of defining pK_a is the pH at which the concentrations of the acid and its conjugate base are equal or the pH at which the acid is half titrated.

$$pH = pK_a + \log 1$$

or

$$pH = pK_a$$

Measurement of pH

pH indicators

Theory An approximate idea of the pH of a solution can be obtained using indicators. These are organic compounds of natural or synthetic origin whose colour is dependent upon the pH of the solution. Indicators are usually weak acids which dissociate in solution.

$$Indicator = Indicator^- + H^+$$

Applying the Henderson–Hasselbalch equation,

$$pH = pK_{In} + \log_{10} \frac{[Indicator^-]}{[Indicator]}$$

The two forms of the indicator have different colours and, as can be seen from above, the actual colour of the solution will depend upon the pK_{In} and the pH. The greatest colour change occurs around the pK_{In} and this is where the indicator is most useful. For example, if a solution has a pH close to 6 then bromocresol purple with a pK_{In} of 6.2 is the best indicator to use. However, this colour change occurs over a wide pH range, so indicators will only give an approximate indication of pH. Another disadvantage is that indicators are affected by oxidizing agents, reducing agents, salt concentration and protein, so these facts must be borne in mind when using them. A final precaution to be taken when using them is to add only a small quantity of indicator to the solution under examination, otherwise the acid–base equilibrium of the test solution may be displaced and the pH changed.

Indicators are probably of more value in determining the end point of a titration and some of those commonly employed are shown in Table 3.5, which is part of the instructions for an experiment for the determination of pH with indicators (Exp. 3.1).

Accurate measurement of pH

The most convenient and reliable method for measuring pH is by the use of a pH meter which measures the e.m.f. of a concentration cell formed from a reference electrode, the test solution, and a glass electrode sensitive to hydrogen ions (Fig. 3.1).

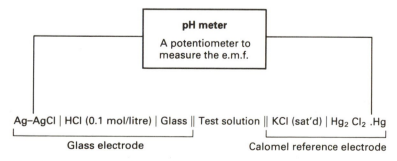

Fig. 3.1   The voltaic cell formed during the measurement of pH

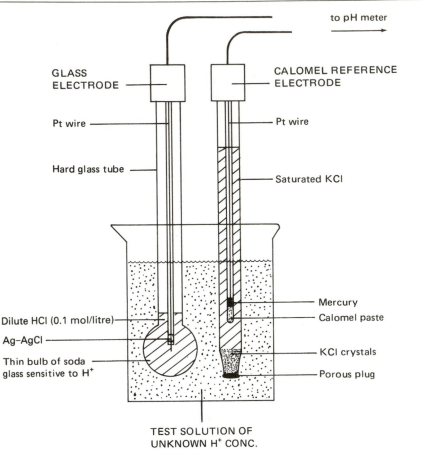

Fig. 3.2   The electrode system for the measurement of pH

Glass electrode   The glass electrode consists of a very thin bulb about 0.1 mm thick blown on to a hard glass tube of high resistance. Inside the bulb is a solution of hydrochloric acid (0.1 mol/litre) connected to a platinum wire via a silver–silver chloride electrode, which is reversible to hydrogen ions (Fig. 3.2). A potential is developed across the thin glass of the bulb which depends on the pH of the solution in which it is immersed. This potential is not readily affected by salts, protein, or oxidizing and reducing agents, so the electrode can be used in a wide variety of media.

The glass electrode in the test solution constitutes a half cell and the measuring circuit is completed by a reference electrode which is not sensitive to hydrogen ions.

Calomel electrode   The reference electrode commonly used is the calomel electrode similar to that illustrated (Fig. 3.2). The calomel electrode is stable, easily prepared, and the potential with respect to the standard hydrogen electrode is accurately known.

pH meter   The e.m.f. of the complete cell (E) formed by the linking of these two electrodes is therefore:

$$E = E_{ref} - E_{glass}$$

where E_{ref} is the potential of the calomel reference electrode (which at normal room temperature is $+0.250$ V) and E_{glass} is the potential of the glass electrode which depends on the pH of the solution under test (pH_s).

$$E_{glass} = 0.342 - 0.058 \, pH_s$$

So

$$E = 0.250 - (0.342 - 0.058 \, pH_s)$$

$$= -0.092 + 0.058 \, pH_s$$

This relationship can be changed by the presence of other potentials in the cell, but these are usually constant and can be allowed for when calibrating the instrument.

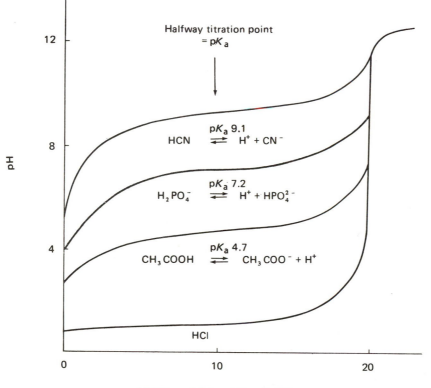

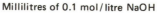

Fig. 3.3   pH titration curves of 20 ml solutions of some 0.1 mol/litre acids

The glass electrode has a very high resistance ($10^6 - 10^8\,\Omega$), so a potentiometer of high input impedence is needed to measure the potential. Some general instructions on the care and operation of a pH meter are given later in this chapter.

Today most pH meters have the glass and reference electrodes combined in one unit and not as separate entities as shown in Fig. 3.2.

Titration curves

When a strong base is mixed with a solution of an acid and the pH measured, a plot of the base added against pH recorded can be obtained and this is known as a titration curve. Several of these curves are shown in Fig. 3.3 and they all have the same shape with the exception of the strong acid HCl.

Similar curves are obtained when a base is titrated with strong acid.

Strong acid and a strong base

There is little change in pH on adding the base until complete neutralization, when only a slight excess of base causes a large increase in pH. In effect the strong acid resists a change in pH or acts as a buffer solution until close to the neutralization point. This can be seen when HCl is titrated with NaOH (Fig. 3.3).

Weak acid and a strong base

All titration curves of a weak acid or base titrated with a strong base or acid are of the same type, since one has a buffer solution present whose pH changes according to the Henderson–Hasselbalch equation. The pK_a values are different for each acid, but the general shape of the curve is the same in all cases.

Buffer solutions

Theory

A buffer solution is one that resists pH change on the addition of acid or alkali. Such solutions are used in many biochemical experiments where the pH needs to be accurately controlled.

From the Henderson–Hasselbalch equation, the pH of a buffer solution depends on two factors; one is the pK_a value and the other the ratio of salt to acid. This ratio is considered to be the same as the amount of salt and acid mixed together over the pH range 4–10, where the concentration of hydrogen and hydroxyl ions is very low and can be ignored. Let us take as an example acetate buffers consisting of a mixture of acetic acid and sodium acetate:

$$CH_3COOH \rightleftharpoons CH_3COO^- + H^+$$

$$CH_3COONa \longrightarrow CH_3COO^- + Na^+$$

Since acetic acid is only weakly dissociated, the concentration of acetic acid is almost the same as the amount put in the mixture; likewise the concentration of acetate ion can be considered to be the same as the concentration of sodium acetate placed in the mixture since the salt is completely dissociated.

Example 1   What is the pH of a mixture of 5 ml of 0.1 mol/litre sodium acetate and 4 ml of 0.1 mol/litre acetic acid?

$$\text{Concentration of CH}_3\text{COO}^- = \tfrac{5}{9} \times 0.1 \text{ mol/litre}$$

$$\text{Concentration of CH}_3\text{COOH} = \tfrac{4}{9} \times 0.1 \text{ mol/litre}$$

$$\text{p}K_a \text{ of acetic acid at 25°C} = 4.76$$

Therefore,

$$\text{pH} = 4.76 + \log \tfrac{5}{4}$$

$$= 4.76 + (+0.097)$$

$$= 4.86$$

Example 2   How is the pH changed on adding 1 ml of 0.1 mol/litre HCl to the above mixture?

Addition of HCl provides H^+ which combines with the acetate ion to give acetic acid. This reduces the amount of acetate ion present and increases the quantity of undissociated acetic acid, leading to an alteration in the salt/acid ratio and hence to a change in pH.

$$\text{Concentration of CH}_3\text{COO}^- = \tfrac{5}{10} \times 0.1 - \tfrac{1}{10} \times 0.1 = 0.04 \text{ mol/litre}$$

$$\text{Concentration of CH}_3\text{COOH} = \tfrac{4}{10} \times 0.1 + \tfrac{1}{10} \times 0.1 = 0.05 \text{ mol/litre}$$

$$\text{p}K_a = 4.76$$

Therefore,

$$\text{pH} = 4.76 + \log \frac{0.04}{0.05}$$

$$= 4.76 + (-0.097)$$

$$= 4.66$$

The pH of the solution has been reduced from 4.86 to 4.66, a change of only 0.2 of a unit, whereas if the HCl had been added to distilled water, the pH would be 2. The solution has, therefore, acted as a buffer by resisting pH change on the addition of acid.

Buffer value β   Buffer solutions vary in the extent to which they resist pH changes, and in order to compare different buffer solutions, Van Slyke introduced the term *buffer value*. When acid or alkali is added to a buffer solution, a titration curve is obtained (Fig. 3.3). The slope of this curve is given by $dB/d(\text{pH})$, where dB is the increment of strong acid or base added in mol/litre and $d(\text{pH})$ the change in the pH increment. This slope is the buffer value $β$ which is always positive since dB is negative when acid is added causing a negative change in pH. As can be seen from the titration curves illustrated (Fig. 3.3), the buffer value is a maximum at the pK_a.

Buffers used in biochemical experiments

Common laboratory buffers

Some of the more common laboratory buffer solutions are shown in Table 3.3. The useful range of these and most other buffers is about 1 unit either side of the pK_a value.

The actual buffer chosen for a particular experiment needs to be selected with care as experimental results may sometimes be due to specific ion effects and not the pH. For example, borate forms complexes with sugars and citrate readily combines with calcium.

Table 3.3 Buffers used in biological investigations

Acid or base	pK_a	pK_a	pK_a
Phosphoric acid	2.1	7.2	12.3
Citric acid	3.1	4.8	5.4
Carbonic acid	6.4	10.3	—
Glycyl glycine	3.1	8.1	—
Acetic acid	4.8	—	—
Barbituric acid	3.4	—	—
Tris*	8.3	—	—
HEPES†	7.6	—	—

* Tris = N-tris(hydroxymethyl)aminomethane.
† HEPES = N-2-Hydroxyethylpiperazine-N'-2-ethanesulphonic acid.

The pH in many biological experiments often needs to be kept constant in the range pH 6–8 but there are few weak acids or bases that are effective buffers at this part of the pH scale and particularly around pH 7.4, the pH of blood.

Bicarbonate   This buffer spontaneously liberates carbon dioxide and must therefore be maintained in an atmosphere of CO_2. Bicarbonate has a pK_a of 6.1 so the buffering capacity around pH 7.4 is poor.

Phosphate   This is probably the most popular buffer but phosphate readily forms complexes with heavy metals. Phosphate is also inconvenient as it plays an active part in a number of biochemical reactions where it can act as an activator, inhibitor or metabolite. The buffering capacity above pH 7.5 is also poor.

Tris   This is a popular buffer since it can be used with heavy metals but it also acts as an inhibitor in some biochemical systems. Tris has a high lipid solubility and penetrates membranes which can be a disadvantage. However, its greatest problem is the temperature effect, which is often overlooked. A tris buffer of pH 7.8 at room temperature has a pH of 8.4 at 4°C and 7.4 at 37°C, so that the hydrogen ion concentration increases ten-fold from the preparation of material at 4°C to its measurement at 37°C. Tris also has a poor buffering capacity below pH 7.5.

Zwitterionic buffers   These buffers were introduced by Good and coworkers to overcome the disadvantages of the traditional materials. As the name suggests they contain both negative and positive groups so do not readily penetrate membranes. They are now widely used in many biological experiments and HEPES has proved to be particularly useful.

$$HO-CH_2-CH_2-\overset{+}{N}H \underset{\diagdown}{\overset{\diagup}{\bigcirc}} N-CH_2-CH_2-SO_3^-$$

N-2-Hydroxyethylpiperazine-*N'*-2-ethanesulphonic acid

(HEPES) $pK_a = 7.6$

pH and life

Most cells can only function within very narrow limits of pH and require buffer systems to resist the changes in pH that would otherwise occur in metabolism. The three main buffer systems in living material are protein, bicarbonate and phosphate and the relative importance of each depends on the type of cell and organism.

Animals

The most important buffering system in mammalian plasma is bicarbonate:

$$H_2CO_3 \rightleftharpoons HCO_3^- + H^+$$

From the Henderson–Hasselbalch equation,

$$pH = pK_a + \log_{10} \frac{[HCO_3^-]}{[H_2CO_3]}$$

The plasma pH, therefore, depends on the ratio of bicarbonate to carbonic acid and not on the absolute concentrations. Any tendency for the pH to change is buffered and can be corrected by adjusting this ratio. For example, large quantities of acids formed during normal metabolism react with bicarbonate to form the weakly dissociated carbonic acid so that free hydrogen ions are effectively mopped up. At the same time, carbonic acid is removed at the lungs as carbon dioxide, thus maintaining the pH of the plasma. The kidneys also play an important role in the maintenance of acid–base balance by adjusting the excretion of acid or base in the urine, so that the pH of urine can normally vary from 4.8 to 7.5 in man.

The extracellular fluid is usually slightly alkaline at pH 7.4 while the average cytoplasmic pH of many animal cells is about pH 6.8 which at 37°C is neutral (Table 3.2). It is almost certain that the pH in the organelles will differ from 6.8 and the pH at most membrane surfaces will be lower than this due to the absorption of H^+ on the negatively charged surface.

The importance of pH control is nicely illustrated in the case of human blood, which must be kept within narrow limits for life and even narrower limits for health (Fig. 3.4).

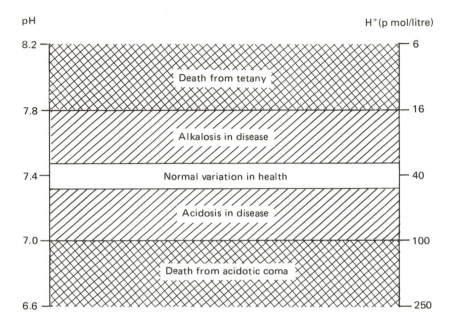

Fig. 3.4   The pH limits of human blood plasma

Plants

The average cytoplasmic pH of plants is similar to that of animals, but the cell sap of the majority of land plants is acid and lies within the range pH 5.2–6.5. Some plant juices are very acid (orange and grapefruit, pH 3), but the acid seems to be present in vacuoles and therefore segregated from the rest of the cytoplasm.

Bacteria

The internal pH of those bacteria examined again appears to be round about neutrality, although many forms grow well at pH 6 or 9, a 1000-fold difference in hydrogen ion concentration. Thiobacilli will actually grow at pH 0, which is that of 1 mol/litre hydrochloric acid, while some fungi grow at pH 11. Most organisms pathogenic to man have, as might be expected, an optimum rate of growth at pH 7.2–7.6.

Clearly, the simpler forms of life appear to be able to survive a wide pH range of the external medium although it does not follow that the internal pH varies to any extent.

Practical exercises

The care and use of the pH meter

The precise method of operation of a pH meter depends on the particular model, so detailed instructions here are inappropriate. However, some general guidance on the use and care of the electrode assembly will be useful.

1. *New glass electrode* New electrodes must be soaked in 0.1 mol/litre HCl or distilled water for several hours before use.
2. *Mixing* The solution must be thoroughly stirred before measuring the pH and this is ideally carried out with a stream of pure nitrogen, although a magnetic stirrer is probably most convenient for class work.
3. *Temperature* The beaker should be thermostatted since K_w, pK and the pH of the standard solution all vary with temperature. The temperature compensator on the instrument does not allow for the above factors, but only for the change in e.m.f. with temperature of the electrode assembly.
4. *Electrode contamination* The electrodes should be washed in distilled water before and after use and must not be touched. In particular they should be thoroughly washed after measuring the pH of a solution with a high concentration of biological macromolecules as these may adhere to the glass and distort subsequent pH measurements unless removed.
5. *Standardization* The pH meter is calibrated before use by means of a standard solution. In the United Kingdom, potassium hydrogen phthalate is the recommended standard; at 15°C a 0.05 mol/litre solution has a pH of 4.0. The meter should be calibrated with a solution whose pH is close to that under test and several convenient standards are given in Table 3.4.
6. *Measurements at extreme pH values* The glass electrode remains accurate down to quite low pH values, but in strong alkali the recorded pH is often too low due to interference by other ions, particularly sodium. When lithium glass is used, this error is considerably reduced, but ordinary glass electrodes cannot be used above pH 10. When titrating at these high pH values it is best to use KOH and not NaOH, and it is also advisable to use a burette with a soda lime trap to prevent absorption of CO_2.
7. *Storage* The pH meter should be switched to zero but the mains switch left on. After use, the electrodes are stored in distilled water and must never be allowed to

Table 3.4 Primary standards for the calibration of a pH meter

	pH	
	25°C	37°C
1. 0.05 mol/litre Potassium hydrogen phthalate	4.01	4.02
2. 0.025 mol/litre Potassium dihydrogen phosphate 0.025 mol/litre Disodium hydrogen phosphate	6.86	6.84
3. 0.01 mol/litre Sodium tetraborate	9.18	9.06

dry out. If this does occur, then the electrodes need to be soaked in water and calibrated frequently before their measurements can be relied on.

Titration curves

Practical limits of titration curves   If 0.1 mol/litre strong acid or base is used in a titration, the curves will asymptotically approach pH 1 or pH 13 after complete neutralization. Likewise, the limits for 0.01 mol/litre solutions will be pH 2 or pH 12 so that pK_a values below 2 or above 12 cannot be determined using 0.01 mol/litre solutions.

Solvent correction   Experimental titration curves must be corrected for the amount of acid or base consumed in titrating the solvent, usually distilled water. This is carried out as follows:

1. Plot the titration curves for the same volume of sample and water.
2. Select a pH value on the curve for the sample and note the volume of acid or base used; let this be X ml. Likewise, note the amount of acid or base consumed in order to bring the water to the same pH value; let this be Y ml.
3. The actual amount of acid consumed in the titration of the sample is, therefore, given by $(X - Y)$ ml. Repeat this for a number of pH values and plot the corrected titration curve.

Determination of pK_a pK_a values can be obtained from titration data by three methods:

1. The pH at the point of inflection is the pK_a value and this may be read directly. A more convenient way is to plot $dB/d(pH)$, the buffer value, against pH when a maximum is obtained at the pK_a.
2. By definition, the pK_a value is equal to the pH at which the acid is half titrated. The pK_a can, therefore, be obtained from a knowledge of the end point of the titration.
3. The ratio of salt/acid can be calculated from the experimental data and a graph prepared of log_{10} salt/acid against pH. The intercept on the axis is the pK_a value.

Table 3.5   Colour change and useful pH range of some common indicators

Indicator	Colour of indicator		pK_a	Useful pH range
	Acid	Base		
Thymol blue	Red	Yellow	1.7	1.2– 2.8
Bromophenol blue	Yellow	Blue	4.0	3.0– 5.0
Methyl red	Red	Yellow	5.0	4.3– 6.1
Bromocresol purple	Yellow	Purple	6.3	5.5– 7.0
Phenol red	Yellow	Red	7.9	6.8– 8.2
Phenolphthalein	Colourless	Red	9.7	8.3–10.0

Experiments

Experiment 3.1   The determination of pH using indicators

MATERIALS

1. Indicators (solutions in aqueous ethanol of the indicators in Table 3.5: these are available commercially)

$\underline{100}$

100 ml

2. Samples to be tested (saliva, orange juice, lemonade, egg white) diluted 1 in 10

1 litre

METHOD

Pipette 1–2 ml of each sample into a test tube, add two drops of methyl red indicator solution, and observe the colour. Repeat the exercise with other indicators until the approximate pH is found by comparing the colours of the sample with those of the acidic, intermediate, and basic forms of the indicator (Table 3.5).

Pipette 2 ml of laboratory distilled water into a test tube and determine the pH using the indicators as described.

Why is the pH much lower than 7?

What happens to the pH when expired air is blown through the water? Explain your observations.

Experiment 3.2   Titration of a mixture of a strong and a weak acid

PRINCIPLE

Gastric juice consists of a mixture of free HCl and weak acids such as phosphate, protein and organic acids. A quantitative assay of both types of acidity can be of value in clinical diagnosis. The relative amounts of HCl and weak acids are determined by titration with an indicator that has two pH ranges such as thymol blue. The pH ranges and the accompanying colour changes for thymol blue are:

pH 1.2–2.8   Red–Orange–Yellow,
pH 8.0–9.6   Yellow–Green–Blue.

Thus, titration to the first colour change gives a measure of the amount of free HCl, while titration to pH 9.6 gives the total acidity.

If genuine gastric juice is not readily available, then a mock sample can be prepared by mixing hydrochloric and acetic acids.

MATERIALS

1. Gastric juice or mock sample prepared by mixing 40 ml of 0.1 mol/litre hydrochloric acid, 15 ml of 0.1 mol/litre acetic acid, and 45 ml of water

$\underline{100}$

2.5 litres

2 Potassium hydroxide (0.1 mol/litre)

2.5 litres

3. Thymol blue (0.1 per cent w/v in 20 per cent v/v ethanol)

100 ml

4. Microburette 10 ml

100

METHOD

Place 10 ml of sample in a small conical flask, add four drops of thymol blue indicator and titrate with 0.1 mol/litre potassium hydroxide until the reddish–orange colour changes to orange–yellow; carefully note the volume of potassium hydroxide used. Continue the titration until the full blue colour forms and again note the volume of alkali required.

Calculate the 'free HCl' and total acidity of the sample as ml 0.1 mol/litre acid/100 ml of sample.

Normal values for genuine gastric juice are:

Free HCl 20–30 ml 0.1 mol/litre acid/100 ml

Total acidity 30–70 ml 0.1 mol/litre acid/100 ml

Compare the results obtained on the mock sample with those expected on theoretical grounds.

Experiment 3.3 Titration of a strong acid with a strong base

MATERIALS	10
1. Hydrochloric acid (0.1 mol/litre)	500 ml
2. Potassium hydroxide (0.1 mol/litre)	500 ml
3. Burette 25 ml	10
4. pH meter	5

METHOD

Pipette 20 ml of 0.1 mol/litre HCl into a 100 ml beaker and measure the pH. Titrate this solution with 0.1 mol/litre KOH and measure the pH after the addition of 1 ml aliquots of alkali. When nearing the end point, reduce the increments of alkali added to give a reasonable change in pH.

Plot the titration curve and compare it with that calculated from theory as previously described.

Experiment 3.4 Titration of a weak acid with a strong base

MATERIALS	10
1. Acetic acid (0.1 mol/litre and 0.01 mol/litre)	500 ml
2. Potassium hydroxide (0.1 mol/litre and 0.01 mol/litre)	1 litre
3. Burette 25 ml	10
4. pH meter	5

METHOD

Titrate 20 ml of 0.1 mol/litre acetic acid as above. Plot the titration curve and deduce the pK value using the methods given. Repeat the titration using 0.01 mol/litre acetic acid and compare both plots with the theoretical curve obtained from the relationship:

$$pH = pK + \log_{10} \frac{[Salt]}{[Acid]}$$

What are the differences between these curves and why?

Experiment 3.5   The determination of pK_a

MATERIALS

1. Solutions of the following acids (1.4 g/litre):

$\underline{10}$
250 ml
(assuming
two acids
per student)

Acid	Mol. wt	pK
Acetic	60.05	4.8
Benzoic	122.12	4.2
Hippuric	179.17	3.6
Imidazole	68.08	7.0
Lactic	90.08	3.9
p-Nitrophenol	139.11	7.1

It is suggested that these acids be identified by a code number known only to the demonstrator.

2. Potassium hydroxide (0.02 mol/litre)   1 litre
3. Burette 25 ml   10
4. pH meter   5

METHOD

Titrate the unknown solution with the standard alkali and deduce its equivalent weight. Add half the amount needed for complete neutralization to another portion of the acid and measure the pH of the solution; this is the pK of the acid. From the data, identify the acid as far as possible using the above list.

Experiment 3.6 pK_a values of a dicarboxylic acid

MATERIALS

1. Dicarboxylic acid solution (0.1 mol/litre)

$\underline{10}$
250 ml
(assuming
two acids
per student)

Acid	pK_1	pK_2
Malic	3.5	5.0
Oxalic	1.3	4.3
Succinic	4.2	5.6
Tartaric	3.0	4.4

(As in the last experiment, the identity of the acid is not known to the student)

2. Potassium hydroxide (0.1 mol/litre)   1 litre
3. Burette 25 ml   10
4. pH meter   5

METHOD
Prepare a titration curve of the acid provided and identify it from the pK values obtained.

Experiment 3.7   Acetate buffers

MATERIALS
1. Sodium acetate (0.1 mol/litre)
2. Acetic acid (0.1 mol/litre)
3. Hydrochloric acid (0.1 mol/litre)
4. pH meter

METHOD
Mix the solutions of sodium acetate and acetic acid as given in the table and record the pH. Calculate what the pH should be from the Henderson–Hasselbalch equation using the examples given earlier.

Add a further 2 ml of 0.1 mol/litre HCl to each mixture, measure the pH again and compare this value with that obtained by calculation.

Tube no.	1	2	3	4	5	6
Sodium acetate (ml)	2	6	10	14	18	18 ml of
Acetic acid (ml)	16	12	8	4	0	water

From your results explain why some solutions are better buffers than others. The pK_a of acetic acid at 25°C is 4.76.

Experiment 3.8   Titration curves of amino acids

PRINCIPLE

Amino acids are present as zwitterions at neutral pH and are amphoteric molecules that can be titrated with both acid and alkali.

$$^+H_3N-\underset{\underset{\text{Acid}}{|}}{\overset{\overset{R}{|}}{CH}}-COOH \underset{}{\overset{\pm H^+}{\rightleftharpoons}} {}^+H_3N-\underset{\underset{\text{Neutral}}{|}}{\overset{\overset{R}{|}}{CH}}-COO^- \underset{}{\overset{\pm H^+}{\rightleftharpoons}} H_2N-\underset{\underset{\text{Alkali}}{|}}{\overset{\overset{R}{|}}{CH}}-COO^-$$

The strong positive charge on the $-NH_3^+$ group induces a tendency for the $-COOH$ to lose a proton, so amino acids are strong acids. Some amino acids have other ionizable groups in their side chains and these can also be titrated.

MATERIALS
1. Hydrochloric acid (100 mmol/litre)
2. Sodium hydroxide (100 mmol/litre)
3. Amino acids (100 mmol/litre glycine, alanine, histidine, and lysine: 50 mmol/litre glutamic acid)
4. pH meter
5. Burette (10 ml)

<u>10</u>
250 ml
250 ml
250 ml

5
10

METHODS

Pipette 10 ml of the amino acid solution into a 100 ml beaker. Standardize the pH meter and determine the pH of the solution. Add the 100 mmol/litre HCl from a burette in small amounts at first and determine the pH after each addition. Continue adding the acid in larger quantities until the pH falls to about 1.3.

Wash the electrodes in distilled water, restandardize the pH meter, and titrate a further 10 ml with 100 mmol/litre NaOH solution to pH 12.5.

Determine the pK values from your curves and compare them with the values given below.

Amino acid	pK_1 $-COOH$	pK_2 $-NH_2$	pK_3 side chain (R)
Glycine	2.4	9.7	—
Alanine	2.3	9.9	—
Glutamic acid	2.2	9.7	4.3 -carboxyl
Histidine	1.8	9.2	6.0 -imidazole
Lysine	2.2	9.0	10.5 -amino

pH calculations

1. Calculate the pH of (a) 0.1 mol/litre HCl, (b) 0.1 mol/litre KOH, (c) 10 mmol/litre HCl, (d) 50 mmol/litre H_2SO_4.

2. What are the pK_a values of succinic acid, given that the acid dissociation constants are $K_1 = 6.4 \times 10^{-5}$ and $K_2 = 2.7 \times 10^{-6}$.

3. Calculate the hydrogen ion concentration of blood plasma, pH 7.42. What will be the pH in acidosis if this hydrogen ion concentration is doubled?

4. A solution is prepared by dissolving sodium benzoate and benzoic acid in water to give a 9 mmol/litre concentration of each compound. The mixture has a pH of 4.21 at 25°C. Calculate the acid dissociation constant of benzoic acid.

5. Calculate the amount of glycine and HCl required to prepare 1 litre of 0.1 mol/litre buffer of (a) pH 2.0 and (b) pH 3.0 given that the pK_a value of the glycine carboxyl group is 2.4.

Further reading

Chang, R., *Physical Chemistry with Application to Biological Systems*, 2nd edn. Clarendon Press, Oxford, 1979.

Gardner, M. L. G., *Medical Acid–Base Balance, The Basic Principles*. Cassell, London, 1978.

Price, N. C. and Dwek, R. A., *Principles and Problems in Physical Chemistry for Biochemists*. Clarendon Press, Oxford, 1979.

Van Holde, K. E., *Physical Biochemistry*, 2nd edn. Prentice-Hall, Englewood Cliffs, 1985.

The Developing Role of the Zwitterionic Buffer, BDH Product Information, Poole, Dorset.

SECTION TWO   Chemical and physical techniques

4. Chromatography

General introduction

Principles of chromatography

Chromatographic techniques employ mild conditions and separate molecules on the basis of differences of size, shape, mass, charge, solubility and adsorption properties. The term *chromatography* is derived from the Greek meaning 'coloured writing' and was first used by the Russian botanist Tswett to describe the separation of coloured plant pigments on a column of alumina. There are many different types of chromatography but they all involve interactions between three components: the mixture to be separated, a solid phase and a solvent (Fig. 4.1; Table 4.1). The

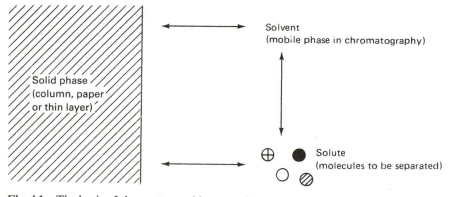

Fig. 4.1   The basis of chromatographic separations

Table 4.1   Solute property used for chromatographic separations

Technique	Solute property	Solid phase	Solvent
Gel filtration	Size and shape	Hydrated gel	Usually aqueous
Adsorption chromatography	Adsorption	Adsorbent usually inorganic material	Non-polar
Partition chromatography	Solubility	Inert support	Mixture of polar and non-polar solvents
Ion exchange chromatography	Ionization	Matrix with ionized groups	Aqueous buffer

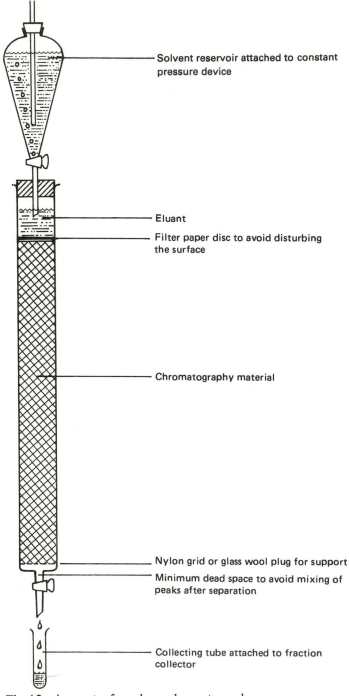

Fig. 4.2 Apparatus for column chromatography

magnitude of these interactions depends on the particular method and the solute–solid interaction is dominant in ion exchange chromatography whereas the solute–solvent interaction is more important in partition chromatography.

A brief introduction is given to the theory and practice of these techniques but not all compounds behave exactly as predicted on theoretical grounds and the best conditions for a particular separation often have to be found by practice.

Chromatographic separations can be carried out on a column of material and this is the method used for the isolation and preparation of compounds. Alternatively, separations can be made in one or two dimensions on paper or a thin layer and these techniques are used for analytical work.

Column chromatography

Separation of compounds by column chromatography must be one of the most widely used techniques in biochemical work. It is therefore appropriate to consider some of the general precautions to be taken when preparing and running columns before dealing with the various types of chromatographic separations.

Columns   Chromatography columns are usually glass and, generally, long columns give good resolution of components but wide columns are better for dealing with large quantities of material. The essential features of a chromatography column are shown in Fig. 4.2.

Preparation of the material   A wide range of materials are used in chromatographic separations and all need to be equilibrated with the solvent before preparing the column. In addition, some form of pretreatment is often required; for example, some gel filtration materials need to be swollen, adsorbents need to be 'activated' by heating or acid treatment, and ion exchange resins have to be obtained in the required ionized form by washing.

During the equilibration with solvent the material is allowed to settle and the fine particles remaining in suspension are removed by decantation (Fig. 4.3(a)). If this is not carried out, the flow rate of solvent down the column will be considerably reduced due to clogging by these fine particles.

The pouring of the column   The chromatography column is packed with material by filling it about one-third full with solvent and slowly adding a slurry of the material in the solvent. This is carefully poured down a glass rod, as shown, to stop air bubbles being trapped in the column (Fig. 4.3(b)). The suspension is allowed to settle and excess solvent run off. This process is repeated until the column is the required height. The column is then washed thoroughly with solvent and the level of the liquid kept just above the surface of the material.

Application of the sample   The sample is first dissolved in the solvent or dialysed against the eluting buffer before loading it on to the column. In most class experiments the concentrated sample is carefully pipetted on to the surface and the tap opened until

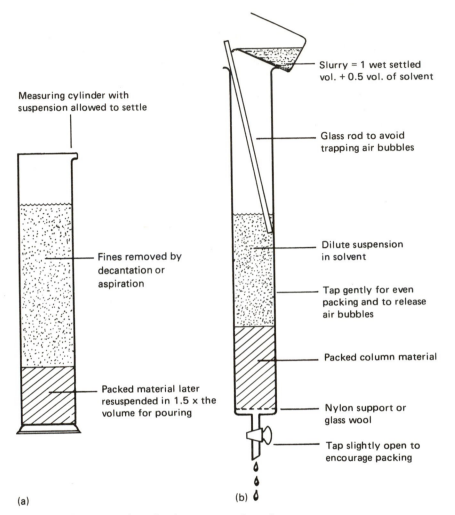

Measuring cylinder with suspension allowed to settle

Fines removed by decantation or aspiration

Packed material later resuspended in 1.5 x the volume for pouring

Slurry = 1 wet settled vol. + 0.5 vol. of solvent

Glass rod to avoid trapping air bubbles

Dilute suspension in solvent

Tap gently for even packing and to release air bubbles

Packed column material

Nylon support or glass wool

Tap slightly open to encourage packing

(a)   (b)

Fig. 4.3   The preparation of a chromatography column

the top of the column is just below the level of the meniscus. The solvent reservoir is connected, and a constant head of liquid maintained at the top of the column from a pressure reservoir.

Elution   The next stage is to remove the materials from the column in order by eluting with an appropriate solvent. In *displacement development*, the solvent interacts more strongly with the chromatographic material than the compound on the column, thus displacing the bound molecules. Large quantities of material can be separated in this way since about 50 per cent of the total column capacity is used. The separation is adequate but for better resolution of peaks *elution development* is preferred. In this case, no more than 10 per cent of the total capacity is loaded on to the column. The solvent interacts with the column more weakly than the solute molecules and overrides the bound molecules, gradually eluting them from the column. This is probably the most commonly used means of elution and different molecules are removed from the column by changing the strength or pH of the eluting solvent in a stepwise fashion or by means of a gradient which can be linear, concave or convex.

The collection and analysis of fractions   The effluent from the column is collected into a series of test tubes, either manually or with a fraction collector. Each fraction is then analysed for the presence of the compounds being examined and an *elution profile* prepared of the amount eluted against the effluent volume.

Paper chromatography

Principle   Cellulose in the form of paper sheets makes an ideal support medium where water is adsorbed between the cellulose fibres and forms a stationary hydrophilic phase.

The mixture is spotted on to the paper, dried and the chromatogram developed by allowing the solvent to flow along the sheet. The solvent front is marked and, after drying the paper, the positions of the compounds present in the mixture are visualized by a suitable staining reaction. The ratio of the distance moved by a compound to that moved by the solvent is known as the R_f *value* and is more or less constant for a particular compound, solvent system and paper under carefully controlled conditions of solute concentration, temperature and pH (Fig. 4.4).

The R_f is related to the partition coefficient α:

$$\alpha = \frac{A_1}{A_s}\left(\frac{1}{R_f} - 1\right)$$

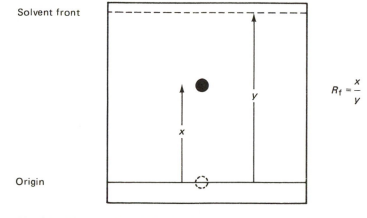

Fig. 4.4   The meaning of R_f value in paper chromatography

A_l = area of cross-section of liquid phase, A_s = area of cross-section of solid stationary phase.

For an homologous series, a plot of R_m against the homologue number gives a straight line.

$$R_m = \log\left(\frac{1}{R_f} - 1\right)$$

Exceptions to these equations are found when the compounds to be separated interact with the supposedly inert cellulose support.

Preparation of the sample   Biological materials should be desalted before chromatography by electrolysis or electrodialysis. Excess salt results in a poor chromatogram with spreading of spots and changes in their R_f values. It can also affect the chemical reactions used to detect the compounds being separated. Macromolecules such as proteins are also removed prior to chromatography by ultrafiltration or gel filtration. The sample (10–20 µl) is then applied to the paper with a micropipette.

Paper   Whatman No. 1 is the paper most frequently used for analytical purposes. Whatman No. 3 MM is a thick paper and is best employed for separating large quantities of material; the resolution is, however, inferior to Whatman No. 1. For a rapid separation, Whatman Nos 4 and 5 are convenient, although the spots are less well defined. In all cases, the flow rate is faster in the 'machine direction', which is normally noted on the box containing the papers. The paper may be impregnated with a buffer solution before use or chemically modified by acetylation. Ion exchange papers are also available commercially.

For the separation of lipids and similar hydrophobic molecules, silica-impregnated papers are available commercially.

Solvent   This choice, like that of the paper, is largely empirical and will depend on the mixture investigated. If the compounds move close to the solvent front in solvent A then they are too soluble, while if they are crowded around the origin in solvent B then they are not sufficiently soluble. A suitable solvent for separation would, therefore, be an appropriate mixture of A and B, so that the R_f values of the components of the mixture are spread across the length of the paper. The pH may also be important in a particular separation, and many solvents contain acetic acid or ammonia to create a strongly acidic or basic environment.

Ascending chromatography   The sheet of paper is supported on a frame with the bottom edge in contact with a trough filled with solvent. Alternatively, the paper can be rolled into a cylinder, fastened with a paper clip and stood in the solvent. The arrangement is contained in an airtight tank lined with paper saturated with the solvent to provide a constant atmosphere and separations are carried out in a constant temperature room (Fig. 4.5).

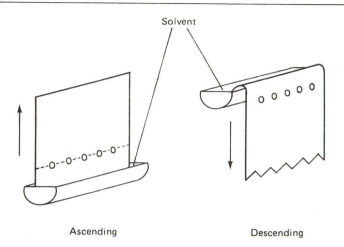

Fig. 4.5   Ascending and descending paper chromatography

Descending chromatography   This method is convenient for compounds which have similar R_f values since the solvent drips off the bottom of the paper, thus giving a wider separation. In this case, of course, the R_f values cannot be measured, and substances

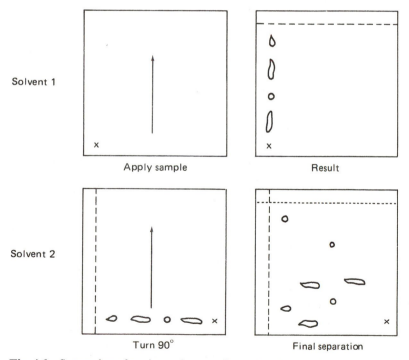

Fig. 4.6   Separation of a mixture by two-dimensional paper chromatography

are compared with a standard reference compound such as glucose, for example in the case of sugars:

$$R_g = \frac{\text{Distance moved by compound x}}{\text{Distance moved by glucose}}$$

Two-dimensional chromatography   The mixture is separated in the first solvent, which should be volatile: then after drying, the paper is turned through 90° and separation is carried out in the second solvent. After location, a map is obtained and compounds can be identified by comparing their position with a map of known compounds developed under the same conditions (Fig. 4.6).

Detection of spots   Most compounds are colourless and are visualized by specific reagents. The location reagent is applied by spraying the paper or rapidly dipping it in a solution of the reagent in a volatile solvent. Viewing under ultraviolet light is also useful since some compounds which absorb strongly show up as dark spots against the fluorescent background of the paper. Other compounds show a characteristic fluorescence under ultraviolet light.

Thin layer chromatography

Principle   Separation of compounds on a thin layer is similar in many ways to paper chromatography, but has the added advantage that a variety of supporting media can be used so that separation can be by adsorption, ion exchange, partition chromatography, or gel filtration depending on the nature of the medium employed. The method is very rapid and many separations can be completed in under an hour. Compounds can be detected at a lower concentration than on paper as the spots are very compact. Furthermore, separated compounds can be detected by corrosive sprays and elevated temperatures with some thin layer materials, which of course is not possible with paper.

Production of thin layer   The R_f value is affected by the thickness of the layer below 200 μm and a depth of 250 μm is suitable for most separations. There are several good spreaders on the market which, when carefully used, can produce an even layer of required thickness. Calcium sulphate is sometimes incorporated into the adsorbent to bind the layer to the plate and, because of this, it is advisable to work rapidly once the adsorbent is mixed with water. There are now a number of prepared thin layer plates that are available commercially and these may be more convenient to use than trying to prepare plates in the laboratory.

Development   It is essential to make sure that the atmosphere of the separation chamber is fully saturated, otherwise R_f values will vary widely from tank to tank. This can be ensured by using as small a tank as possible and lining the walls with paper soaked in the solvent. Development of the plate is usually by the ascending technique and is very rapid.

Location   The compounds are located as for paper chromatography by spraying with the appropriate reagent or by scanning, in the case of radioactive substances.

Experiments

Adsorption chromatography

THEORY

Compounds are adsorbed on to the column and an equilibrium set up between molecules bound to the column and those free in solution. The extent of the binding is governed by the charge, van der Waals' forces, dipole interactions, hydrogen bonding and steric factors, and depends on the structure of the compounds. The mass of solute adsorbed per unit weight of adsorbent (m) depends on the concentration of the solute (c) and Langmuir derived an equation on the basis that: (a) only a monolayer is adsorbed, and (b) only a proportion of the molecules in collision will result in adsorption. This is known as the Langmuir adsorption isotherm:

$$m = \frac{K_1 K_2 c}{1 + K_2 c}$$

K_1 is a measure of the number of active adsorption sites per unit weight of adsorbent and depends on the nature of the adsorbent. K_2 is a measure of the affinity of solute for the adsorbent and is affected by all the components of the system.

Langmuir assumed only one binding site, but in practice there are a number of different sites on the surface of the adsorbent, each with a different affinity, thus giving a series of Langmuir-type isotherms. Hinshelwood, therefore, suggested that the equation

$$m = \sum \frac{K_1 K_2 c}{1 + K_2 c}$$

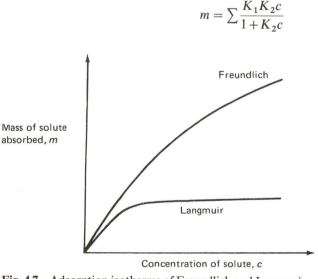

Fig. 4.7   Adsorption isotherms of Freundlich and Langmuir

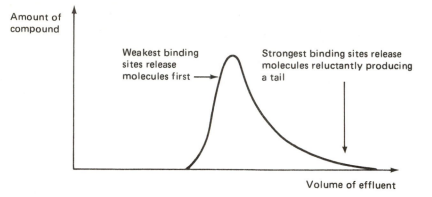

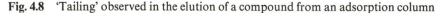

Fig. 4.8 'Tailing' observed in the elution of a compound from an adsorption column

gives a more accurate picture of the situation, and this, indeed, approximates to the Freundlich adsorption isotherm found in practice:

$$m = Kc^x$$

K and x are constants depending on the particular system used.

The difference between these two isotherms is illustrated in the graph (Fig. 4.7).

This mixture of binding sites of differing affinities is the cause of 'tailing' observed in the elution profile of compounds (Fig. 4.8). This tailing can be overcome by eluting with a suitable gradient of pH, ionic strength, or polarity, so that the more strongly adsorbed molecules meet a higher concentration of displacing compound than the more weakly bound molecules.

Experiment 4.1   The separation of leaf pigments by adsorption chromatography

PRINCIPLE

The principle is similar to the previous experiment. Benzene is highly toxic and should be handled with care and in a fume cupboard, not an open laboratory.

MATERIALS	10
1. Chromatography column (20 cm × 1 cm)	5
2. Fresh spinach leaves	—
3. Alumina	100 g
4. Calcium carbonate	200 g
5. Sucrose (finely sieved icing sugar is suitable)	200 g
6. Sodium sulphate (anhydrous)	100 g
7. Petroleum ether (b.p. 60–80°C)	1 litre
8. Methanol	500 ml
9. Benzene (*Care:* toxic!)	500 ml
10. Waring blender	5

METHOD

Extraction   Homogenize the leaves in a Waring blender, then extract by shaking with a mixture of petroleum ether, methanol and benzene (45:15:5). Remove the residue by filtration and wash the filtrate four times with water to remove the methanol. Avoid vigorous shaking or an emulsion will form. Remove the last traces of water by adding anhydrous sodium sulphate, filter to remove the solid, and concentrate the extract to a few millilitres by careful evaporation in a fume chamber.

Preparation of column   Prepare slurries of the column materials in petroleum ether and pack the column with alumina (5 cm), calcium carbonate (7 cm) and sucrose (7 cm), inserting a filter paper disc between each adsorbent. Gentle suction may be applied to the bottom of the column to assist packing. Wash the column with several volumes of the eluting solvent, a mixture of benzene and petroleum ether (1:4).

Separation and elution of the pigments   When the top of the column is almost dry, add the extract and elute with solvent. If the flow rate is too slow, apply gentle pressure to the top of the column. Collect the fractions and plot the absorption spectrum of each coloured peak.

Ion exchange chromatography

THEORY

The matrix   The earliest ion exchange materials were based on synthetic resins of an aromatic nature and were suitable for the separation of inorganic ions and small molecules. However, they could not be used for the separation of large molecules such

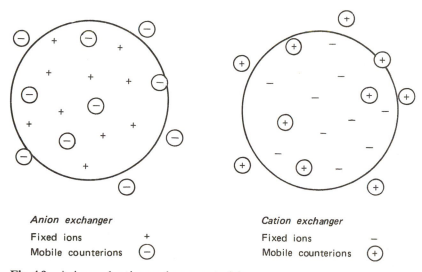

Anion exchanger

Fixed ions   +
Mobile counterions   ⊖

Cation exchanger

Fixed ions   −
Mobile counterions   ⊕

Fig. 4.9   Anion and cation exchange materials

as proteins which cannot penetrate the closely linked structure of the resin and tend to be denatured by the hydrophobic matrix.

The first ion exchangers suitable for macromolecules were based on cellulose but they had low capacities as too much substitution made the cellulose soluble. Since then other materials have been developed based on dextran and an acrylamide type of polymer which have a much higher capacity.

Ionizable groups   Charged groups are attached to the matrix and the type of group defines the nature and strength of the ion exchanger. These groups may be either *anionic* or *cationic*, according to the nature of their affinity for either negative or positive ions. For example, the cation exchange materials exchange positive ions, so it is the charge carried by the exchangeable ion which decides whether a material is anionic or cationic and not the charge carried on the matrix (Fig. 4.9).

These two types can be further divided into materials that contain *strongly ionized* groups, such as $-SO_3H$ and $-\overset{+}{N}R_3$, and the *weakly ionized* groups, such as $-COOH$, $-OH$ and $-NH_2$. The strong ion exchange resins are completely ionized and exist in the charged form except at extreme pH values:

$$-SO_3H \rightleftharpoons -SO_3^- + H^+$$

$$-NR_3OH \rightleftharpoons -\overset{+}{N}R_3 + OH^-$$

The weak ion exchange materials, on the other hand, contain groups whose ionization is dependent on the pH, and they can only be used at maximum capacity over a narrow pH range.

$$-COOH \rightleftharpoons -COO^- + H^+$$

$$-NH_3^+ \rightleftharpoons -NH_2 + H^+$$

Table 4.2   Functional groups of some ion exchangers

Anion exchangers		Structure	Type
Quaternary aminoethyl	QAE	$-CH_2 \cdot CH_2 - \overset{+}{N} \overset{\diagup C_2H_5}{\underset{\diagdown C_2H_5}{-CH_2 - CH(OH) - CH_3}}$	Strong
Diethylaminoethyl	DEAE	$-CH_2 \cdot CH_2 - \overset{+}{N}H \overset{\diagup C_2H_5}{\underset{\diagdown C_2H_5}{}}$	Weak

Cation exchangers		Structure	Type
Sulphopropyl	SP	$-CH_2 \cdot CH_2 \cdot CH_2 - SO_3^-$	Strong
Phospho	P	$-PO_4H_2^-$	Intermediate
Carboxymethyl	CM	$-CH_2COO^-$	Weak

As a rough guide, resins containing carboxyl groups have a maximum capacity above pH 6, while those with amino groups are effective below pH 6.

The number of ionizable groups determines the capacity of the ion exchanger. The *total capacity* is the number of ionizable groups per gram of material, whereas the *available capacity* is the amount of a given molecule that can bind under defined experimental conditions. In the case of some materials, large molecules may be unable to penetrate the matrix and can only react with charged groups on the surface. In this case, the available capacity will be considerably less than the total capacity.

The ionizable groups commonly met are given in Table 4.2 together with their abbreviations.

Ion exchange equilibria   The typical way that an ion exchange material functions is illustrated in the following example, where an anion exchange resin containing amino groups is used to separate two negatively charged ions X^- and Y^- (Fig. 4.10).

Although ion exchange materials are claimed to be monofunctional, in that only the ion exchange process is used in separation, in practice some molecular sieving and

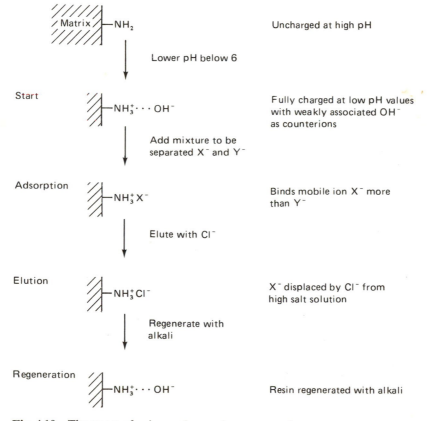

Fig. 4.10   The stages of anion exchange chromatography

adsorption can occur. The adsorption is small, but it can sometimes be used to separate closely related compounds.

Elution of bound ions   The bound ions can be removed by changing the pH of the buffer. For example, as the pH of a protein moves towards its isoelectric point, the net charge decreases and the macromolecule is no longer bound. Separation is achieved as other charged proteins remain on the column. Alternatively, ions can be removed by increasing the ionic strength, when high concentrations of ions in the solvent displace the bound ions by increasing the competition for the charged groups of the ion exchange material. The pH or ionic strength can be altered sharply, by changing the eluting buffer, or, gradually, by means of a gradient as discussed earlier.

It is important to realize that the binding of ions is not an all or none phenomenon, but involves an equilibrium between those ions firmly bound to the materials and those free in solution. The extent of the binding depends on the nature of the resin and the temperature, ionic strength and composition of the solvent. In the case of weakly ionized resins, the uptake of ions also depends on the degree of ionization of the resin and the materials to be separated.

Preparation of material   Ion exchange materials are first allowed to swell in the buffer and the fines removed as described earlier.

The ion exchange material is then obtained in the required ionic form by washing with the appropriate solution. For example, the H^+ form of a cation exchange resin is obtained by washing the material with hydrochloric acid then water until the washings are neutral. Similarly the Na^+ form is prepared by washing the resin with sodium chloride or sodium hydroxide then water as above.

The final stage before preparing the column is to equilibrate the material by stirring with the eluting buffer.

Experiment 4.2   The separation of amino acids by ion exchange chromatography

MATERIALS	10
1. Chromatography column (20 cm × 1.5 cm)	5
2. Strongly acidic resin	30 g
3. Hydrochloric acid (4 mol/litre)	1 litre
4. Hydrochloric acid (0.1 mol/litre)	4 litres
5. Glass wool	—
6. Amino acid mixture. (Dissolve aspartic acid, histidine and lysine in 0.1 mol/litre HCl to a final concentration of 2 mg/ml)	10 mg of each
7. Tris–HCl buffer (0.2 mol/litre, pH 8.5)	3 litres
8. Sodium hydroxide (0.1 mol/litre)	2 litres
9. Separating funnels (500 ml)	10
10. Acetate buffer (4 mol/litre, pH 5.5)	250 ml
11. Ninhydrin reagent. (Dissolve 20 g of ninhydrin and 3 g of hydrindantin in 750 ml of methyl cellosolve and add 250 ml of acetate buffer.) Prepare fresh and store in a brown bottle	1 litre

12. Methyl cellosolve (ethylene glycol monomethyl ether)   1 litre
13. Ethanol (50 per cent v/v)   1 litre
14. Ninhydrin (2 g/litre in acetone). (*Care*: carcinogenic!)   100 ml
15. Oven at 105°C   1

METHOD

Preparation of the column   Gently stir the resin with 4 mol/litre HCl until fully swollen (15–30 ml/g dry resin). Allow the resin to settle, then decant the acid. Repeat the washing with 0.1 mol/litre HCl, resuspend in this solution, and prepare the column as previously described.

Elution of amino acids   Carefully apply 0.2 ml of the amino acid mixture to the top of the column, open the tap and allow the sample to flow into the resin. Add 0.2 ml of 0.1 mol/litre HCl, allow to flow into the column as before, and repeat the process twice. Finally, apply 2 ml of 0.1 mol/litre HCl to the top of the resin and connect the column to a reservoir containing 500 ml of 0.1 mol/litre HCl. Adjust the height of the reservoir to give a flow rate of about 1 ml/min and collect a total of forty 2 ml fractions. Test five of the tubes at a time for the presence of amino acids by spotting a sample from each tube on to a filter paper: dip this in the acetone solution of ninhydrin and heat in an oven at 105°C. If amino acids are present, they will show up as blue spots on the filter paper. When the first amino acid has been eluted, remove the reservoir of 0.1 mol/litre HCl and allow the level of acid to fall to just above the resin. Run 2 ml of 0.2 mol/litre tris–HCl buffer (pH 8.5) on to the column, then connect to a reservoir of this buffer and continue with the elution until the third amino acid is removed from the column.

Detection of amino acids   Adjust the pH of each tube to 5 by the addition of a few drops of acid or alkali. Add 2 ml of the buffered ninhydrin reagent and heat in a boiling water bath for 15 min. Cool the tubes to room temperature, add 3 ml of 50 per cent v/v ethanol and read the extinction at 570 nm after allowing the tubes to stand for 10 min.

Table 4.3   The pK values and pH of zero net charge of three amino acids

Amino acid	pK_1	pK_2	pK_3	Isoionic point
Aspartic acid	2.0	3.9	10.0	2.9
Histidine	1.8	6.0	9.2	7.6
Lysine	2.2	9.0	10.5	9.7

Set up the appropriate blanks and standards, and plot the amount of amino acid in each fraction against the volume eluted.

    In what order are the amino acids eluted and why? (Table 4.3)

Experiment 4.3   The separation of proteins by ion exchange chromatography

PRINCIPLE

The binding and release of compounds to ion exchange materials is dependent on pH and salt concentration. This is illustrated in the following experiment involving the fractionation of proteins from human serum.

At pH 7, most of the serum proteins are bound to DEAE–cellulose apart from the γ-globulins which appear in the first fraction. The bound proteins can then be eluted in three more fractions by a stepwise increase in the ionic strength of the eluting medium.

MATERIALS
		$\underline{10}$
1. Ion exchange column of DEAE–cellulose (3 cm × 1 cm)		5
2. Eluting buffer: sodium phosphate (0.01 mol/litre; pH 7.0)		250 ml
3. Eluting buffer containing 0.1 mol/litre NaCl		50 ml
4. Eluting buffer containing 0.2 mol/litre NaCl		50 ml
5. Eluting buffer containing 0.3 mol/litre NaCl		50 ml
6. Human serum		7 ml
7. UV spectrophotometers		5

METHOD

Wash the ion exchange column thoroughly with the elution buffer then carefully pipette 0.5 ml of serum on top of the column. Allow the serum to pass into the DEAE–cellulose then elute slowly with 3 ml of the eluting buffer. Collect the eluate in a test tube and repeat the elution with the buffer containing increasing concentrations of salt.

Fraction	NaCl (mol/litre)	Volume (ml)
1	0	3
2	0.1	3
3	0.2	3
4	0.3	3

Recovery of protein   Calculate the recovery of protein by determining the protein content of the serum and the four fractions using the extinction at 280 nm.

Characterization of protein   Identify the proteins in the serum and each of the fractions using cellulose acetate electrophoresis (Exp. 5.2). From these results, explain the relative ease of binding of the proteins to the DEAE–cellulose.

Partition chromatography

THEORY

The separation of compounds that are readily soluble in organic liquids, but sparingly soluble in water, is carried out by *adsorption chromatography*, while ionizable water-soluble compounds are best separated by *ion exchange chromatography. Partition chromatography* is intermediate between adsorption and ion exchange chromatography, and compounds that are soluble in both water and organic solvents are readily separated by partition methods.

When a compound is shaken with two immiscible solvents it will distribute itself unevenly between the two phases and, at equilibrium, the ratio of the concentration of the compound in the two solvents is constant and is known as the partition coefficient (α).

$$\alpha = \frac{\text{Concentration of } x \text{ in solvent 1}}{\text{Concentration of } x \text{ in solvent 2}}$$

In partition chromatography one solvent, usually water, is held on the stationary supporting phase which is in the form of a column of film of inert material. The other phase consists of a mobile, water-saturated, organic liquid that flows past the stationary phase. The components of a mixture are separated if their partition coefficients between the solvents are sufficiently different (Fig. 4.11).

Experiment 4.4   The separation of amino acids by two-dimensional paper chromatography

PRINCIPLE

A two-dimensional map is prepared of standard amino acids which will be needed in subsequent experiments with amino acids and proteins. Ninhydrin reacts with all α-amino acids to give a purple colour. Other compounds also react if present, and these include primary and secondary aliphatic amines and some non-aromatic heterocyclic nitrogen compounds. The imino acids, proline and hydroxyproline, react to give a yellow colour.

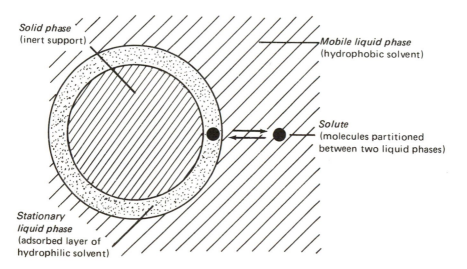

Fig. 4.11   The principles of liquid–liquid partition chromatography

MATERIALS   10
1. Equipment for two-dimensional chromatography   5
2. Whatman No. 1 chromatography paper (20 cm × 20 cm)   30
3. Solvent 1 (butanol:glacial acetic acid:water, 12:3:5)   4 litres
4. Solvent 2 (phenol–water). Add 125 ml of water to a 500 g bottle of   2500 g
    phenol, replace the stopper and allow to stand overnight. *Care*:
    phenol can give nasty burns on the skin. Just before use, add a few
    drops of 0.88 ammonia to the solvent and mix well
5. Ninhydrin location reagent. (Dissolve 0.2 g in 100 ml of acetone just   —
    before use)
6. Oven at 105°C   5
7. Standard amino acids. Prepare small volumes of 10 g/litre solutions   10 ml
    in 10 per cent v/v isopropanol. Sometimes a drop of acid or alkali is
    needed to bring the compound into solution. Alanine, aspartic acid,
    cysteine HCl, cystine, glutamic acid, glycine, histidine HCl,
    hydroxyproline, leucine, isoleucine, lysine HCl, methionine,
    ornithine HCl, phenylalanine, proline, serine, threonine,
    tryptophan, tyrosine, valine

METHOD

Examine the amino acid map provided (Fig. 4.12). Select four amino acids whose R_f values differ widely, and spot 20 μl of each one on to the corner of a sheet of Whatman

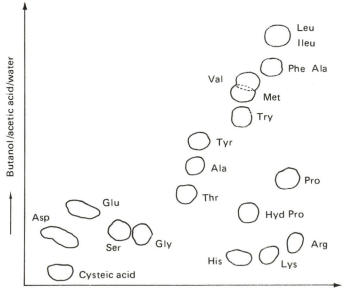

Fig. 4.12 Separation of a mixture of amino acids in two dimensions on Whatman No. 1 paper

No. 1 chromatography paper. Dry the spot in a current of air and mount the sheet on the metal frame. Repeat this until all the amino acids are accounted for and finally spot a mixture of all the amino acids on to a sheet of chromatography paper. Each amino acid should be incorporated into two separate mixtures in order to confirm the identity.

Place the frame with one of the edges of the paper to which the sample spot is adjacent into the first solvent and allow to run for most of the day. In the evening, remove the frame, place it in the fume chamber, and dry the paper in a current of cold air. Arrange the frame so that the second edge to which the spot was adjacent now dips in the second solvent and run the chromatogram overnight (Fig. 4.6). Dry the papers as previously and remove them from the frame. Rapidly dip each paper through the ninhydrin reagent and hang in the fume chamber to allow the acetone to evaporate. Develop the colours by heating at 105°C for 2–3 min. Construct an amino acid map and compare it with Fig. 4.12.

Experiment 4.5   The identification of sugars in fruit juices using thin layer chromatography

MATERIALS	10
1. Thin layer plates of silica gel G. (Prepare a slurry of silica gel G in 0.02 mol/litre sodium acetate and pour the plates 250 μm thick. Activate before use by heating at 105°C for 30 min)	10
2. Separation chambers	5
3. Solvent (ethyl acetate:isopropanol:water:pyridine, 26:14:7:2)	5 × tank volume
4. Fruit juices from fresh fruit (lemon, orange, grapefruit, pineapple)	5 ml
5. Absolute ethanol	100 ml
6. Oven at 105°C	—
7. Hair drier	3
8. Spray guns	5
9. Standard sugar solutions (10 g/litre in 10 per cent v/v isopropanol) of glucose, fructose, xylose, ribose, lactose, galactose, chamnose	5
10. Aniline–diphenylamine location reagent. (Prepare fresh by mixing 5 volumes of 10 g/litre aniline and 5 volumes of 10 g/litre diphenylamine in acetone with 1 volume of 85 per cent phosphoric acid. *Care*: toxic!)	—

METHOD

Add 3 ml of ethanol to 1 ml of fruit juice and centrifuge to remove denatured protein. Carefully spot the supernatant on to a thin layer plate together with some standard sugar solutions. Place the plate in a chamber saturated with solvent and develop the chromatogram until the solvent front is close to the top of the plate. Draw a line across the plate at this point and remove the chromatogram when the solvent reaches the mark. Dry the plate in a stream of cold air and locate the sugars by spraying the plates with freshly prepared aniline–diphenylamine in a fume chamber and heating briefly at 100°C. Note the colour of each sugar and use this and the R_g value to identify the sugars present in the fruit juices.

Experiment 4.6   The separation of lipids by thin layer chromatography

PRINCIPLE

Lipids in biological material are present as a complex mixture and are first fractionated into a number of groups by solvent extraction. Resolution of the compounds within each group can then be carried out by thin layer chromatography. The solvent is selected according to the lipids under investigation, but, generally, neutral lipids are separated with non-polar solvents and charged lipids with polar solvents. A number of supporting media can be used for lipid fractionation and the actual choice of material depends on the group of lipids and solvent employed. Silica gel has been widely used and separation with this material is on the basis of adsorption as well as partition.

In the following experiments, the lipids are separated into groups according to their polarity and examples from each group are included as representative standards. Some naturally occurring oils and fats are examined for the presence of the classes of lipid considered.

MATERIALS	10
1. Thin layer plates of silica gel G	10
2. Separation chambers	5
3. Solvent (petroleum ether, b.p. 60–70°C:diethylether:glacial acetic acid; 80:20:1)	5 × tank capacity
4. Hydrocarbons (*n*-hexadecane and *n*-octadecane)	5 ml
5. Cholesterol esters (cholesterol acetate, cholesterol oleate, and cholesterol stearate)	5 ml
6. Vitamin A ester (vitamin A palmitate)	5 ml
7. Triacylglycerols (glycerol trioleate, glycerol tristearate, and glycerol tripalmitate)	5 ml
8. Free fatty acids (oleic acid, palmitic acid and stearic acid)	5 ml
9. Sterols (cholesterol)	5 ml
10. Naturally occurring oils (olive oil and cod liver oil)	5 ml
11. 2′,7′-Dichlorofluorescein (2 g/litre in 95 per cent v/v ethanol)	200 ml
12. Sulphuric acid (50 per cent v/v)	200 ml
13. Oven at 110°C	3
14. Ultraviolet lamp	3

METHOD

Clean the glass plates with ethanol, then pour an aqueous slurry of the gel on to their surface 250 μm thick. Activate the plates by heating at 110°C for 1 h and allow to cool. Spot out 20–50 μl of an approximately 1 per cent w/v solution of each lipid in the solvent and develop the chromatogram. Locate the lipids by spraying with the dichlorofluorescein solution and view the plates in ultraviolet light (270 nm). The lipids show up as green fluorescent spots against a dark background. Alternatively, the spots can be visualized by spraying the plates with 50 per cent v/v sulphuric acid followed by heating in the oven at 110°C for 10 min. Be careful with the sulphuric acid spray!

In what order do the lipids separate and why?

Gel filtration

Molecules can also be separated on the basis of differences in their size by passing them down a column containing swollen particles of a gel. Small molecules can enter the gel but larger molecules are excluded from the cross-linked network (Fig. 4.13).

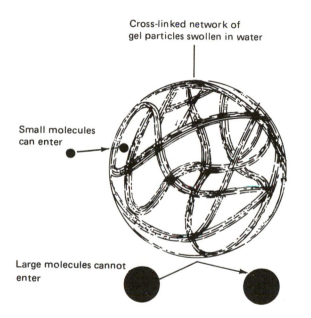

Fig. 4.13 The principle of gel filtration

This means that the accessible volume of solvent is very much less for molecules totally excluded from the gel than for small molecules which are free to penetrate. The separation of molecules by gel filtration is illustrated in the diagram (Fig. 4.14). First the mixture of large (●) and small (·) molecules is placed on top of the column (Fig. 4.14(a)). As they pass down the column, the small molecules diffuse into the gel and follow a longer path than the large molecules, which are completely excluded from the gel particles (Fig. 4.14(b)). Eventually, complete separation occurs, with the large molecules leaving the column first and the smaller ones last (Fig. 4.14(c)).

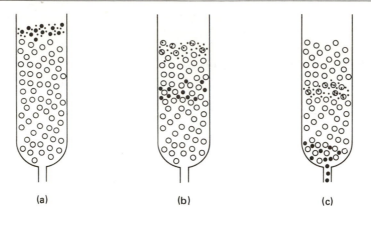

(a)   (b)   (c)

o Gel particles
• Large molecules
· Small molecules

Fig. 4.14   The principle of gel filtration

THEORY

The total volume of a column of gel (V_T) is the sum of the volume of the gel matrix (V_g), the volume of water inside the gel particles (V_i) and the volume of water outside the gel grains (V_o):

$$V_T = V_o + V_i + V_g$$

V_o is known as the void volume, which is the volume of liquid required to elute compounds that are completely excluded from the gel grains.

V_i can be calculated from a knowledge of the dry weight of the gel (a) and the water regain (W_r):

$$V_i = aW_r$$

The elution volume (V_e) of a compound is the volume required to elute that compound from a column:

$$V_e = V_o + K_d V_i$$

K_d indicates the fraction of the inner volume accessible to a particular compound and is independent of the geometry of the column:

$$K_d = \frac{V_e - V_o}{V_i} = \frac{V_e - V_o}{aW_r}$$

If a molecule is completely excluded from the gel, then $K_d = 0$ and $V_e = V_o$; while if a molecule has complete accessibility to the gel, then $K_d = 1$ and $V_e = V_o + V_i$. Molecules, therefore, normally have K_d values between 0 and 1. If K_d is greater than 1, then adsorption of the compound on the gel has occurred.

Table 4.4 Properties of some Sephadex gels

Type	Molecular weight fractionation range		Water regain (g/g dry gel)	Bed volume (ml/g dry gel)
	Polysaccharides	Peptides and proteins		
G10	up to 700	up to 700	1.0	2–3
G15	up to 1 500	up to 1 500	1.5	2.5–3.5
G25*	100– 5 000	1 000– 5 000	2.5	4–6
G50	500– 10 000	1 500– 30 000	5.0	9–11
G75	1 000– 50 000	3 000– 80 000	7.5	12–15
G100	1 000–100 000	4 000–150 000	10.0	15–20
G150	1 000–150 000	5 000–300 000	15.0	20–30
G200	1 000–200 000	5 000–600 000	20.0	30–40

* G25 → 200 are available as superfine grade also. By permission of Pharmacia.

MATERIALS

Pharmacia manufacture cross-linked dextrans (Sephadex) and agarose (Sepharose), and cross-linked agarose (Sepharose CL) and Sephacryl, while Bio-Rad Laboratories make polyacrylamide (Biogel P), agarose (Biogel A), porous glass (Bio-Glas), and polystyrene (Bio-Beads). The degree of cross-linking, etc., is carefully controlled to give a range of products able to fractionate molecules over a limited size range. This is illustrated in the case of Sephadex (Table 4.4) and Sepharose (Table 4.5). The useful fractionation range of molecules is only approximate since separation depends on the shape and, to a minor extent, the charge, as well as the size of the molecules. Sephadex G25 and G50 are made in several particle sizes: fine, coarse, medium and superfine grades. The fine particles are suitable for most laboratory work requiring high resolution, while the coarse material is convenient for preparative work with large columns since this gives a higher flow rate.

Table 4.5 Types of Sepharose

Type	Agarose conc. (%)	Exclusion limits (mol. wt $\times 10^6$)	
		Polysaccharides	Proteins
2B + 2B − CL	2	20	40
4B + 4B − CL	4	5	20
6B + 6B − CL	6	1	4

By permission of Pharmacia.

Sepharose is stable from pH 4 to pH 10 and can be used over the temperature range 0–30°C, and Sepharose CL is stable over pH 3–14 and can be used up to temperatures of 70°C. These more porous gels are used to fractionate very large molecules such as nucleic acids and viruses.

After use the columns can be washed in water and stored in the cold room for quite a time provided a trace of preservative such as phenol or chloroform is added to prevent bacterial growth.

Experiment 4.7   The separation of blue dextran and cobalt chloride on Sephadex G25

PRINCIPLE

This experiment illustrates how gel filtration separates molecules by reason of differences in their size. Blue dextran is a polymer with a molecular weight of several million and is therefore excluded from the gel ($K_d = 0$). Cobalt chloride, on the other hand, with a low molecular weight is freely accessible to the gel particles ($K_d = 1$). The compounds are both coloured so the progress of the filtration can be followed by observing the separation of the coloured bands. The completed fractionation is then analysed by measuring the extinction of each fraction at 625 nm and 510 nm (λ_{max} blue dextran = 625 nm; λ_{max} $CoCl_2 \cdot 6H_2O$ = 510 nm).

MATERIALS	$\underline{10}$
1. Sephadex G25	20 g
2. Chromatography column (20 cm × 1 cm)	5
3. Sodium chloride (0.154 mol/litre)	1 litre
4. Blue dextran in saline	
5. Cobalt chloride in saline	
6. Colorimeters	5

METHOD

Suspend 4 g of Sephadex G25 in the sodium chloride solution and leave to swell for 3 h. During this time, stir the solution and remove any 'fines' by decantation. Prepare a column of the gel (18 cm × 1 cm) by pouring the gel suspension into the column and allowing it to settle under gravity while maintaining a slow flow rate through the column. Use a large volume of solvent relative to the amount of the gel to avoid trapping air bubbles in the column.

Add 0.5 ml of the mixture of blue dextran and cobalt chloride to the top of the column and elute with the isotonic saline. Collect 2 ml fractions until all the cobalt chloride has been eluted. Measure the extinction of each fraction at 510 nm and 625 nm and plot the elution profile.

1. Explain the results obtained!
2. Calculate V_o and V_i.
3. How do these values compare with those expected?

Experiment 4.8   The separation of proteins by gel filtration

PRINCIPLE

Sephacryl is a polymer of allyl dextran covalently cross λ-linked with N,N'-methylenebisacrylamide and can be obtained in a pre-swollen form. The material gives much faster flow rates than Sephadex or Bio-Gel and the S-200 can be used to separate proteins with a molecular weight between 5000 and 250 000.

The molecular weight of a protein can be determined from the elution volume after calibrating the column with known molecular weight markers.

MATERIALS   <u>10</u>

1. Sephacryl S-200 columns (20 cm × 2 cm) equilibrated with the   5
   elution buffer
2. Elution buffer (0.5 mol/litre NaCl in 0.1 mol/litre tris–HCl buffer, pH   3 litres
   8.0)
3. Buffer reservoirs   5
4. Fraction collectors   5
5. Molecular weight standards (5 mg of each protein per ml of elution   3 ml
   buffer)

Standard	Mol. wt	Log mol. wt
cytochrome c	12 400	4.09
ovalbumin	45 000	4.65
bovine serum γ-globulin	210 000	5.32
blue dextran	2 000 000 (approx.)	—

6. Lactate dehydrogenase   1 ml
7. Spectrophotometer   5
8. Reagents for the assay of glucose (Exp. 9.5 or 9.6)   —
9. Reagents for the assay of lactate dehydrogenase (Exp. 12.3)   —

METHOD

Running the column   Open the tap on the column and allow the level of the elution buffer to fall until it is just above the top of the gel. Mix 0.1 ml of lactate dehydrogenase solution and 0.4 ml of the markers and place this on top of the column. Allow the mixture to enter the gel, add a small volume of the elution buffer until the markers are clear of the top then connect the buffer reservoir. Adjust the height of the reservoir to give a flow of about 50 ml/h and collect 3 ml fractions using the fraction collector. Locate the position of the blue dextran and the glucose and examine the fractions between these compounds for the presence of the proteins.

Detection of compounds
Blue dextran: Extinction at 625 nm
γ-Globulin and ovalbumin: Extinction at 280 nm
Cytochrome c: Extinction at 412 nm
Glucose: Exp. 9.5 or 9.6
Lactate dehydrogenase   Exp. 12.3

Molecular weight determination   Calibrate the column by determining the volume at which the blue dextran is eluted (V_o) and the volume when the glucose appears ($V_o + V_i$). Determine the elution volume (V_e) of the three marker proteins and the LDH and calculate the K_d values. Plot a graph of K_d against $\log_{10}$ mol. wt for each marker and calculate the apparent molecular weight of the LDH.

Affinity chromatography

THEORY

Principle   Affinity chromatography is a type of adsorption chromatography in which there is a high degree of specificity in the interaction between the adsorbent and the compound to be separated. In the case of enzymes, the ligand attached to the adsorbent is usually a powerful inhibitor with a high affinity constant which will bind only one enzyme in a complex mixture of proteins.

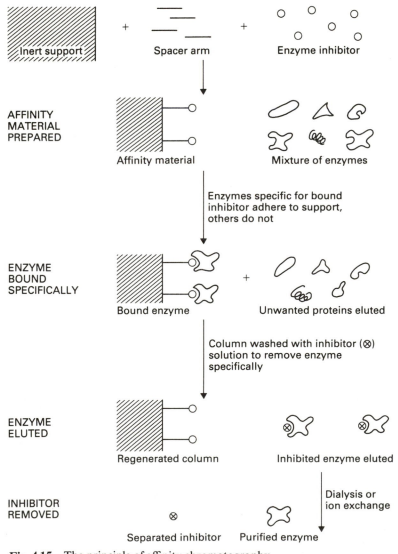

Fig. 4.15   The principle of affinity chromatography

Affinity material   The *ligand* is covalently attached to a supporting medium so that the chromatographic material can be designed for a specific purification task. The *matrix* has to be macroporous to allow large molecules access to the binding sites and needs to have good flow properties. It has to be devoid of non-specific adsorption sites but must contain functional groups to which ligands can be attached. There are a number of commercially available materials that fit these criteria including cross-linked Sepharose (CL-Sepharose) available from Pharmacia.

A *spacer arm* is nearly always inserted between the matrix and the ligand so that large molecules can gain access to the binding sites. If this spacer arm is too short, then steric hindrance can still occur while if it is too long there is an increased risk of non-specific adsorption, particularly of hydrophobic compounds. In practice, a spacer arm of 2–10 C atoms has been found to be optimal.

Purification by affinity chromatography   The principle of affinity chromatography is relatively simple to understand and this is illustrated in Fig. 4.15 which shows the purification of an enzyme from a mixture of other proteins. However, like most things the practice is not quite so straightforward and technical problems arise such as unwanted side reactions during the synthesis and attachment of the ligand to the side arm and matrix. This tends to increase the non-specific adsorption of the affinity material with a consequent lowering of the specificity. In other cases, the binding of the compound to the ligand can be so strong that it becomes quite difficult to remove it from the affinity column.

However, in spite of these and other technical problems, separation by affinity chromatography can lead to a very high degree of purification in a single step.

Experiment 4.9   The preparation of an affinity column

PRINCIPLE

Cyanogen bromide reacts with the free hydroxyl groups of CL-Sepharose 4B to give imidocarbonates which can then be reacted with any compound that contains a primary amino group. In the present experiment the starting material used is CNBr-activated Sepharose 4B which avoids having to handle free cyanogen bromide, an unpleasant and hazardous reagent. The spacer arm is introduced next by reacting the activated Sepharose with diaminobutane, then lengthened by the addition of succinic anhydride so that the ligand (3-amino-*N*-methyl pyridinium ion) can be linked to the activated spacer arm by a peptide bond using a water-soluble carbodiimide. The affinity material prepared is *N*-methyl-3-aminopyridine agarose abbreviated to MAP-agarose for convenience.

MATERIALS

	10
1. CNBr-activated Sepharose 4B	10 g
2. HCl (1 mmol/litre)	3 litres
3. Sodium borate buffer (0.1 mol/litre, pH 9.5)	1 litre
4. Succinic anhydride	50 g

5. NaOH (4 mol/litre)	1 litre
6. Diaminobutane dihydrochloride	150 g
7. 3-Aminopyridine	10 g
8. Iodomethane (*Care*: volatile and possibly carcinogenic!)	20 ml
9. Acetone	250 ml
10. 1-(3-dimethylaminopropyl)-3-ethyl carbodiimide hydrochloride	5 g
11. Sintered glass funnels (G3) to fit Buchner filtration flask	5
12. Mechanical shakers	5
13. Elution buffer (0.5 per cent w/v Triton X-100 in 30 mmol/litre sodium phosphate buffer, pH 7.4)	3 litres
14. Round-bottomed flasks (250 ml)	10
15. Melting point apparatus	2
16. Melting point tubes	—

METHOD

Initial washing and swelling   Weigh out 2 g of the freeze dried CNBr–Sepharose and mix with 1 mmol/litre HCl on a sintered glass filter (G3). Add about 400 ml of solution in several aliquots over 15–20 min and suck off the supernatant between successive additions. The washing of the gel at low pH preserves the ligand binding of the Sepharose better than washing at pH 7.4. The CNBr-activated Sepharose contains dextran and lactose to preserve its activity during freeze drying and these are removed by the washing procedure. During this time, prepare diaminobutane dihydrochloride solution by dissolving 25 g in 100 ml of the borate buffer. Carefully check that the pH is 9 and adjust if necessary.

Addition of the spacer arm   Give the gel a rapid final wash with cold borate buffer, remove the supernatant by suction then immediately add 100 ml of ice-cold diaminobutane dihydrochloride solution. Transfer the mixture to a 250 ml round-bottomed flask and shake gently at 4°C overnight.

$$\text{CNBr-activated Sepharose} \quad -O-C=NH \; + \; H_2N-(CH_2)_4-NH_2$$
$$\underset{OH}{|}$$
$$\downarrow$$
$$\text{Sepharose with side arm} \quad -O-\underset{\underset{NH}{\|}}{C}-NH-(CH_2)_4-NH_2 \quad \text{N-substituted isourea (major product)}$$

Shown as: Sepharose$-NH-(CH_2)_4-NH_2$

Preparation of the ligand   Dissolve 1 g of aminopyridine in 15 ml of acetone and stir with 3 ml of iodomethane for 18 h. Filter the precipitated N-methyl-3-aminopyridinium iodide (MAP) and wash with acetone. Allow the product to dry and record the yield and melting point.

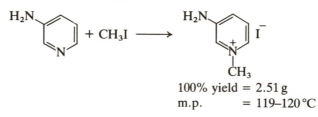

100% yield = 2.51 g
m.p.   = 119–120 °C

Coupling of ligand   Remove the excess amine by filtration on a sintered glass funnel. Wash the Sepharose thoroughly then disperse in 100 ml of water. Transfer the suspension to a flask and cool on ice to 4°C. Prepare a saturated solution of succinic anhydride by adding 10 g to 100 ml of warm water, cool to 4°C, decant if necessary and add to the Sepharose with stirring. Filter the material, wash thoroughly with water to remove the free succinic anhydride, then add 0.13 g of 3-amino-*N*-methyl pyridinium iodide to 6 ml of the succinylated Sepharose in the presence of 0.86 g of 1-(3-dimethylaminopropyl)-3-ethyl carbodiimide hydrochloride. Shake the reaction mixture gently overnight at 4°C and thoroughly wash with the elution buffer before use.

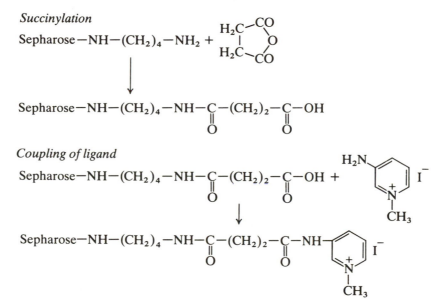

The function of the soluble carbodiimide is to promote the formation of the peptide bond between the —COOH of the spacer arm and the —NH₂ of the ligand. The mechanism is complex but a simplified representation of the role of the carbodiimide in this process is given below.

$$RN{=}C{=}NR + R'COOH + R''NH_2 \longrightarrow R'CONHR'' + RNHCONHR$$

Peptide   Substituted
bond   urea

Suggested timetable

There are several stages where the preparation has to be left overnight and four successive days are needed to carry out this and Exp. 4.10. A convenient timetable is:

Day 1

1. Prepare ligand and leave for 18 h overnight.
2. Wash the CNBr–Sepharose and add the spacer arm, leave overnight.

Day 2

3. Couple ligand to affinity material, leave overnight.
4. Prepare brain extract and store in the deep freeze until day 4.

Day 3

5. Wash MAP–agarose and prepare column.

Day 4

6. Thaw brain extract (assay for AChE and protein) and use affinity column to purify AChE.

Experiment 4.10a   The preparation of rat brain acetylcholinesterase

PRINCIPLE

Extraction   Acetylcholinesterase (AChE) is a membrane-bound enzyme and the first step in any investigation is to bring the enzyme into solution. This can be conveniently achieved by extraction of the tissue with the non-ionic detergent Triton X-100.

The criterion of solubility is an operational one and the enzyme is considered to be soluble if it remains in the supernatant following centrifugation at 100 000 g for 1 h.

MATERIALS	10
1. Rat brains	5
2. Sodium phosphate buffer (30 mmol/litre, pH 7)	100 ml
3. Triton X-100 (1 per cent w/v in 30 mmol/litre sodium phosphate buffer)	100 ml
4. Bladed homogenizers	5
5. Refrigerated ultracentrifuges	5

METHOD

Remove the brain from a rat and wash with the sodium phosphate buffer. Remove the membrane from the organ, finely chop the tissue and homogenize in 10 ml of ice-cold sodium phosphate buffer: add 10 ml of 1 per cent w/v Triton X-100 at 4°C and mix thoroughly with gentle stirring. Centrifuge for 1 h at 100 000 g and carefully remove the supernatant which is the soluble acetylcholinesterase. This can then be stored in the deep freeze without loss of activity until required.

Experiment 4.10b   The assay of acetylcholinesterase

PRINCIPLE

The substrate used in the assay system is acetylthiocholine, the ester of thiocholine and acetic acid. The mercaptan formed as a result of the hydrolysis of the ester then reacts with an oxidizing agent 5,5'-dithiobis-(2-nitrobenzoic acid) (DTNB) which is split into two products, one of which (5-thio-2-nitrobenzoate) absorbs at 412 nm. The activity of the enzyme can thus be measured by following the increase in absorbance at 412 nm in a double-beam spectrophotometer.

Enzyme hydrolysis

$$(CH_3)_3\overset{+}{N}CH_2CH_2SCOCH_3 \xrightarrow{\ H_2O\ } (CH_3)_3\overset{+}{N}CH_2CH_2S^- + CH_3COO^- + 2H^+$$

Production of colour

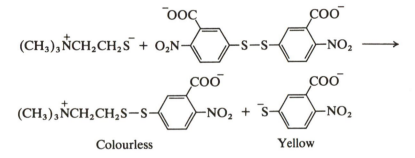

Colourless   Yellow

MATERIALS   10
1. Sodium phosphate buffer (0.1 mol/litre, pH 8)   500 ml
2. DTNB (10 mmol/litre, this is unstable at alkaline pH values and so the   250 mg
   stock solution is prepared just before use by dissolving 39.6 mg in 10 ml
   of sodium phosphate buffer (0.1 mol/litre containing 15 mg sodium
   bicarbonate)
3. Acetylthiocholine iodide (158.5 mmol/litre in the sodium phosphate   10 ml
   buffer, prepare fresh)
4. Recording double-beam spectrophotometers   5

METHOD

Add 50 µl of the enzyme to 3 ml of sodium phosphate buffer (0.1 mol/litre, pH 8) and incubate at room temperature for 5 min. Add 10 µl of DTNB (10 mmol/litre) followed by 20 µl of acetylthiocholine iodide (158.5 mmol/litre) to give a final concentration of 1 mmol/litre of the substrate. Record the increase in absorbance at 412 nm on a double-beam spectrophotometer against a blank of the above mixture prepared at the same time but in the latter case the 50 µl of enzyme is replaced with 50 µl of the buffer solution.

Calculation of enzyme activity   If the change in extinction is ΔE per minute, then the activity in International Units is:

AChE activity $= (\Delta E \times 1000 \times 3.17)/(1.36 \times 10^4 \times 0.05)$
$\Delta E = $ Extinction change per minute
$1000 = $ Factor to obtain µmoles
$3.17 = $ Total volume of reaction mixture (ml)
$0.05 = $ Volume of the enzyme (ml)
$1.36 \times 10^4 = $ Molar extinction coefficient of chromophore at 412 nm (litres mol^{-1} cm^{-1})
AChE activity $= \Delta E \times 4.66$ litre moles min^{-1} ml^{-1}

In some cases, a larger volume of enzyme may be needed and in this case an appropriate adjustment needs to be made to the above equation.

Experiment 4.10c   Purification of acetylcholinesterase by affinity chromatography

PRINCIPLE

Binding   The ligand attached to the agarose by a spacer arm is a strong inhibitor of acetylcholinesterase so that when the crude extract is added to the column, the AChE molecules are specifically bound to the ligand. The other proteins present in the extract do not bind to the affinity material and are removed from the column by washing. This is achieved by running the solution containing the enzyme through the column followed by the eluting buffer until protein is no longer detected in the eluate. If the column is functioning properly, there should be no enzyme detected in any of the fractions collected up to this stage.

Elution   The enzyme is eluted from the column by incorporating an inhibitor (decamethonium bromide) in the elution buffer which has an even higher affinity for AChE than that shown by the ligand.

Reactivation of acetylcholinesterase   This is carried out by adding a cation exchange resin (Amberlite CG-120) which strongly binds the decamethonium bromide.

Removal of protein bound non-specifically   In practice, some non-specific binding of the other enzyme and other proteins always occurs and these are removed before using the column again. This is achieved by eluting with a strong salt solution (1 mol/litre NaCl) which also removes some enzyme activity and by washing with 6 mol/litre guanidine HCl which denatures any remaining proteins and removes them from the affinity material. The column is then given a final wash with the elution buffer and stored in the cold room until required again.

Assay of protein   The detergent Triton X-100 absorbs strongly in the UV at 280 nm and so measurements at this wavelength cannot be used to estimate protein. The

Folin–Lowry method is more sensitive but a precipitate forms in the presence of the detergent and the assay takes some time to complete. The biuret method has a low sensitivity but is probably the best procedure for this experiment when a rapid assay of protein is called for to monitor the fractions. Low extinction values are obtained but they can be read accurately on a spectrophotometer with an external recorder whose sensitivity can be increased.

MATERIALS

<u>10</u>

1. MAP–agarose affinity column (Exp. 4.9) — 5
2. Small chromatography columns (10 cm × 1 cm) — 5
3. Sodium phosphate buffer (30 mmol/litre, pH 7) — 2 litres
4. Triton elution buffer (0.5 per cent Triton X-100 in the sodium phosphate buffer) — 1 litre
5. Decamethonium bromide (10 mmol/litre and 50 mmol/litre dissolved in the Triton elution buffer) — 250 ml
   This is quite toxic and will inhibit your acetylcholinesterase, so be careful!
6. Sodium chloride (1 mol/litre in the Triton elution buffer) — 250 ml
7. Guanidine hydrochloride (6 mol/litre) — 250 ml
8. Amberlite CG-120 — —
9. Reagents for the biuret assay of proteins (Exp. 8.2) — —

METHOD

Thaw the frozen brain extract and remove 1 ml for the assay of protein and AChE. Add the remainder of the solution to the column then wash with the Triton elution buffer collecting 3 ml fractions. Assay each fraction for protein using 0.5 ml in duplicate and continue collecting fractions until protein can no longer be detected in the eluate. Store the remainder of each fraction on ice until required for the assay of AChE.

At this stage elute with 10 mmol/litre decamethonium bromide and collect a further five fractions. Increase the concentration of decamethonium bromide to 50 mmol/litre and collect another five fractions then wash the column with NaCl (1 mol/litre) dissolved in the elution buffer until no further protein or enzyme activity is detected.

Assay each of the fractions for AChE activity and select those fractions that contain the enzyme. Add a small quantity of the ion exchange resin, gently mix and assay these fractions again for enzyme activity.

Finally plot the total protein and AChE in each fraction against the elution volume. Calculate the specific activity of the purified AChE and compare it with that of the crude extract.

How much of the enzyme activity and protein is bound to the column and what are the recoveries on elution?

What effect does the ion exchange resin have on the activity of the eluted AChE?

Further reading

Dean, P. D. G., Johnson, W. S. and Middle, F. A., *Affinity Chromatography a Practical Approach*. IRC Press, Oxford, 1985.

Deyl, Z., *Separation Methods*. Elsevier, Amsterdam, 1984.

Fischer, L., *Gel Filtration Chromatography* (Series: *Laboratory Techniques in Biochemistry and Molecular Biology* T. S. Work and R. H. Burdon eds). Elsevier/North Holland, Amsterdam, 1980.

Smith, I., *Chromatographic and Electrophoretic Techniques*, 4th edn. Vol. 1 *Chromatography*. Heinemann, London, 1976.

Wilson, K., 'Chromatographic Techniques' in *A Biologist's Guide to Principles and Techniques of Practical Biochemistry*, K. Wilson and K. H. Goulding (eds). 3rd edn. Arnold, London, 1986.

5. Electrophoresis

Theory and practice

Electrophoretic mobility

Many biological molecules carry an electrical charge, the magnitude of which depends on the particular molecule and also the pH and composition of the suspending medium. These charged molecules migrate in solution to the electrodes of opposite polarity when an electric field is applied, and this principle is used in electrophoresis to separate molecules of differing charges.

If a molecule of charge q is present in an electric field of strength x then the force on the particle causing it to accelerate is qx. This is balanced by the frictional resistance f to give a terminal velocity v:

$$qx = fv$$

Now assuming Stokes' law for a spherical molecule of radius r moving through a medium of viscosity η, then,

$$f = 6\pi\eta r$$

From these two equations,

$$qx = 6\pi\eta rv$$

The *electrophoretic mobility* $\bar{v}$ of a molecule is defined as the migration per unit field strength, so that,

$$\bar{v} = v/x = q/6\pi\eta r$$

The mobility of the molecules thus depends on the viscosity of the medium (η), and the size, shape (r), and charge on the molecule (q).

The *electrophoretic mobility* depends mainly on the ionizable groups present on the surface of the particle, and the sign and magnitude of the charge carried by the ionizing groups varies according to the ionic strength and pH of the medium in a characteristic manner. Separation of molecules can therefore be effected by selecting the appropriate medium.

Zone electrophoresis

The earliest forms of electrophoresis were carried out in free solution but mixing of components by convection currents proved a problem and this method is no longer used. Today electrophoresis is carried out on a supporting medium which reduces the effect of convection to a minimum. Complete separation of a mixture can be effected into distinct zones, and this technique has replaced the classical method of free boundary electrophoresis. Some commonly used supporting media are now considered.

Paper   Filter paper is cheap and easy to use and is therefore quite suitable for class experiments. It was the earliest material to be used, but has now been largely replaced by other supporting media. The main disadvantage of paper is that some mixing occurs between the zones due to adsorption of the molecules on the cellulose.

Cellulose acetate   This material shows minimal adsorption and clear separation of a mixture into discrete zones. Because of this, compounds are easily eluted with good recovery. Very small quantities of material are needed and the separation can be completed in only an hour or so compared with overnight for paper electrophoresis. Cellulose acetate is, however, more expensive than paper.

Agar-gel   This medium is transparent, which makes it particularly suitable for photometric scanning. Agar is the material of choice for immunoelectrophoresis where it is prepared on microscope slides. Agar is also cheap and readily available.

Starch grains   This medium is prepared by compressing whole starch grains in a buffer into a block. Large molecules are separated and starch-block electrophoresis is ideal for work such as the separation and isolation of isoenzymes.

Polyacrylamide-gel   Polyacrylamide is the most recent material used for electrophoresis and has the big advantage of being transparent so that it can be scanned in the visible and the UV. In addition, even finer resolution of complex mixtures is possible and the pore size can be controlled, so that separation can depend on the size and shape of the molecules as well as the charge.

Buffer solution

The suspension medium needs to be carefully selected since some buffer ions react with the compounds under investigation; borate, for example, forms complexes with sugars while citrate and EDTA will chelate metal ions such as calcium.

The pH chosen depends, of course, on the particular mixture under investigation but generally maximum separation is obtained at the isoelectric point of one of the compounds. The pH selected should not cause chemical changes or denaturation of the molecules under examination.

The ionic strength of the buffer is such that it usually lies in the range 0.05–0.15 and is a compromise between two extremes. At low ionic strength, there is rapid migration and low heat production but marked diffusion. On the other hand, at high ionic strength sharp bands are obtained, but there is a higher heat production and migration is over a shorter distance.

Electrical field

A stable d.c. source is required which gives a constant current or voltage. A field strength from 5 to 8 V/cm is suitable for most separations at room temperature. If the field strength is greater than 10 V/cm then the heating effect is such that an excessive amount of water is lost by evaporation. The buffer then flows from the buffer tanks to replace the water lost by evaporation which causes displacement of the zones. If the heating is excessive, compounds may be denatured. Methods of cooling are available so that higher field strengths can be used but special equipment with efficient cooling and a number of foolproof devices is needed for working with higher voltages. Such specialized high voltage equipment can be used successfully to resolve mixtures of low molecular weight compounds which suffer from excess diffusion as the method is so rapid. However, most common electrophoretic separations involve charged macromolecules where diffusion is minimal.

Electrophoresis apparatus

The electrophoresis apparatus consists of the separating medium connected to two electrode tanks by means of filter paper or gauze pads. The electrodes are divided into two compartments connected by cotton wool wicks; one part contains the platinum electrode and the other is in contact with the electrophoresis medium. Changes in pH

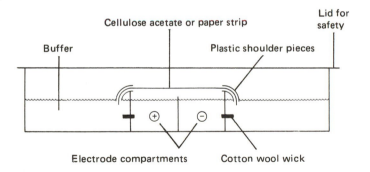

Fig. 5.1 Apparatus for electrophoresis in a horizontal plane

occur in the region of the electrodes even in a buffered solution, and the division into two compartments ensures that such changes close to the support phase are minimized. The connection between this phase and the buffer solution is made by several thicknesses of Whatman 3 MM filter paper or hospital gauze saturated with buffer solution. This connection should be such as to cause a minimum drop in potential across its length. A diagram of a typical tank for horizontal separation is given in Fig. 5.1.

Experiments

Experiment 5.1   The separation of amino acids by paper electrophoresis

PRINCIPLE

The charge carried by a molecule depends on the pH of the medium and this is illustrated in the case of three amino acids: aspartic acid, histidine and lysine. Electrophoresis at low voltage is not usually used to separate low molecular weight compounds because of diffusion, but it is easier to illustrate the relationship between charge and pH with amino acids than with proteins or other macromolecules.

At pH 7.6, histidine will carry zero net charge, aspartic acid will be negatively charged and lysine will be positively charged (Table 4.3).

Amino acid: Histidine   Aspartic acid   Lysine
Charge at pH 7.6: 0   −   +

These three amino acids can therefore be readily separated by electrophoresis. Glucose, an uncharged molecule, is included in the mixture to check for any movement of the origin due to electro-osmosis.

MATERIALS	10
1. Horizontal electrophoresis apparatus	5
2. Power pack	5
3. Amino acids (aspartic acid, histidine, lysine and a mixture of all three in tris–acetate buffer containing 10 g/litre glucose)	10 ml
4. Tris–acetate buffer (0.07 mol/litre, pH 7.6)	5 × tank capacity
5. Ninhydrin location reagent. (Dissolve 0.2 g in 100 ml of acetone just before use)	—
6. Paper strips (10 cm × 2.5 cm)	60
7. Aniline–diphenylamine reagent as in Exp. 4.5	—
8. Citrate buffer (0.07 mol/litre, pH 3.0)	5 × tank volume
9. Oven at 110°C	5

METHOD

Fill both parts of each electrode compartment with buffer solution to the same level; check this by arranging a siphon between them. Remove the siphon, place five filter paper strips as shown (Fig. 5.1), and carefully apply a streak of the amino acid mixture to two of these, avoiding the edge of the paper. Streak three other paper strips with only one amino acid and run all five electrophoresis strips together. Wet the paper from each electrode compartment to within a few centimetres of the point of application, leave the rest to be wetted by capillary attraction, and immediately switch on the current. This way there is minimum spreading of the sample. Carry out electrophoresis for 3 h at 8 V/cm, remove the strips, and dry in an oven at 110°C. Develop one of the strips containing the mixture for glucose (Exp. 4.5) and dip the remaining four strips rapidly in freshly prepared ninhydrin solution: allow the acetone to evaporate in the air and develop the colours by heating in the oven for a few minutes. Identify the amino acids and check for any electro-osmosis.

Repeat the experiment with 0.07 mol/litre citrate buffer, pH 3.0, and explain the results using the data given in Table 4.3.

Experiment 5.2   The separation of serum proteins by electrophoresis on cellulose acetate

PRINCIPLE

There are numerous applications of electrophoresis in medicine and clinical biochemistry and one test frequently carried out is the analysis of serum for changes in the proteins during disease. A rapid and convenient method for this is cellulose acetate electrophoresis illustrated in the following experiment.

MATERIALS $\underline{10}$
1. Horizontal electrophoresis apparatus   5
2. Power pack   5
3. Barbitone buffer (0.07 mol/litre, pH 8.6)   5 × tank capacity
4. Ponceau S protein stain (2 g/litre in 30 g/litre TCA)   2 litres
5. Acetic acid (5 per cent v/v)   2 litres
6. Serum   5 ml
7. Cellulose acetate strips (Oxoid Ltd)   10
8. Whatman 3 MM paper   10
9. Methanol:water (3:2)   50 ml
10. Citrate buffer (0.1 mol/litre, pH 6.8)   20 ml

METHOD

Moisten a strip of cellulose acetate (10 cm × 2.5 cm) by placing it on the surface of the buffer in a flat dish and allow the buffer to soak up from below. Immerse the strip completely by gently rocking the dish, then remove it with forceps. Lightly blot the strip, place in the apparatus and connect it to the buffer compartments with filter paper wicks. Switch on the current and adjust to 0.4 mA per centimetre width of strip. Apply a streak of serum from a microlitre pipette or melting point tube about one-third of the length of the strip from the cathode. This is best carried out by guiding the application with a ruler placed across the tank.

Carry out electrophoresis for $1\frac{1}{2}$–2 h, remove the strips, and stain with Ponceau S for 10 min. Prior heating is not needed here since the TCA fixes the proteins for staining. Remove excess dye from the strip by washing repeatedly in 5 per cent v/v acetic acid. Compare your electrophoresis pattern with that shown in Fig. 5.2.

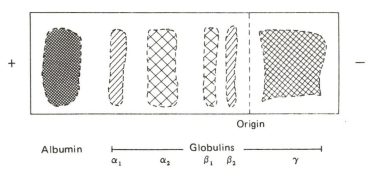

Fig. 5.2   The separation of human serum proteins by electrophoresis on cellulose acetate at pH 8.6

Experiment 5.3   The identification of proteins by electrophoresis following the fractionation of human blood plasma

PRINCIPLE

Plasma contains albumin, fibrinogen and a variety of globulins which can be separated and identified by electrophoresis. These proteins can be fractionated by precipitation with salt or organic solvents and electrophoresis is used to monitor the separation achieved.

Sodium sulphite is used as the salt as this is very effective at room temperature. This salt has the added advantage that it does not interfere with the assay of protein by the biuret or Folin–Lowry method unlike the more commonly used ammonium sulphate. The amount of protein in each fraction can therefore be measured.

Methanol is a convenient organic solvent as this precipitates the globulins with little or no denaturation.

Proteins in human plasma g/100 ml	
Albumin	3.5–4.5
α-Globulins	0.6–1.2
β-Globulins	0.6–1.3
γ-Globulins	0.6–1.5
Total serum proteins	6.0–7.5
Fibrinogen	0.2–0.4

MATERIALS

<table>
<tr><td></td><td></td><td>10</td></tr>
<tr><td>1.</td><td>Sodium sulphite (12.6 g/100 ml)</td><td>100 ml</td></tr>
<tr><td>2.</td><td>Sodium sulphite (15.8 g/100 ml)</td><td>100 ml</td></tr>
<tr><td>3.</td><td>Sodium sulphite (21 g/100 ml)</td><td>100 ml</td></tr>
<tr><td>4.</td><td>Human blood plasma</td><td>2 ml</td></tr>
<tr><td>5.</td><td>Methanol:water (3:2)</td><td>100 ml</td></tr>
<tr><td>6.</td><td>Citrate buffer (0.1 mol/litre, pH 6.8)</td><td>20 ml</td></tr>
<tr><td>7.</td><td>Sodium hydroxide (1 mol/litre)</td><td>100 ml</td></tr>
<tr><td>8.</td><td>Sodium chloride (0.89 g/100 ml)</td><td>100 ml</td></tr>
<tr><td>9.</td><td>Materials for the separation of proteins by electrophoresis (Exp. 5.2)</td><td>—</td></tr>
</table>

METHOD

Salt precipitation   Add 0.5 ml of plasma to 9.5 ml of sodium sulphite solution and leave to stand for 15 min at room temperature after mixing thoroughly. Use all three concentrations of sodium sulphite and collect the precipitates by centrifugation at 3000 g for 10 min. Keep the supernatant from the solution containing 20 per cent sodium sulphite but remove the other supernatants and discard. Turn the centrifuge tubes upside down on filter paper to drain and dissolve the precipitate in a minimum volume of saline. Carry out electrophoresis on the dissolved precipitates and the 20 per cent supernatant and compare the patterns obtained with that of whole plasma.

Organic solvent precipitation   All steps must be carried out at 0°C. Mix 2 ml of plasma with 1 ml of the citrate buffer in a centrifuge tube. Slowly add 7 ml of *ice-cold* methanol in water with stirring. Leave on ice for 30 min, then centrifuge for 10 min at 0°C. Remove the supernatant and dissolve the precipitate in saline. Carry out electrophoresis on these two solutions. Compare this separation with that obtained by salt precipitation.

Assay of protein   Assay the amount of protein in each fraction by repeating the experiment and dissolving the precipitates in 2 ml of 1 mol/litre NaOH. A suitable aliquot is then used to measure the amount of protein present using the biuret method (Exp. 8.2).

Experiment 5.4   Polyacrylamide-gel electrophoresis

PRINCIPLE

Polyacrylamide The gel is prepared by polymerizing acrylamide $(CH_2{=}CH \cdot CO \cdot NH_2)$ and a small quantity of cross-linking reagent, methylenebisacrylamide $(CH_2{=}CH \cdot CO \cdot NH_2)_2 \cdot CH_2$ (bis), in the presence of a catalyst, ammonium persulphate. Tetramethylethylenediamine (TEMED) is also present to initiate and control the polymerization. The gel mixture is allowed to polymerize in small tubes sealed at the bottom with a rubber cap. A layer of water is placed on top of the gel to ensure a flat surface and also to exclude oxygen which inhibits the polymerization.

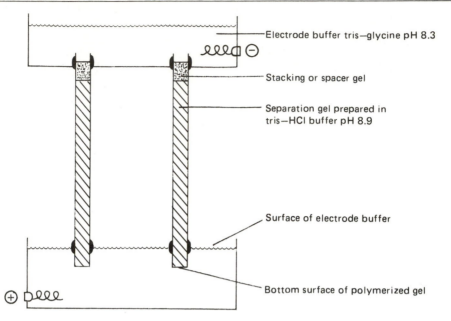

Fig. 5.3   Apparatus for electrophoresis on rods of polyacrylamide

The pore size of the gel can be altered by varying the concentration of monomer in the gel solution. Proteins can be conveniently separated on 7.5 per cent acrylamide, but larger molecules such as ribosomal nucleic acids require a more open gel of 2.5 per cent acrylamide.

The apparatus   A diagrammatic representation of the apparatus used for polyacrylamide-gel electrophoresis on rods is shown in Fig. 5.3: only two rods are shown for the sake of clarity. The stacking gel is of a large pore size to concentrate the sample so that it enters the separation gel as a tight band. The electrophoretic mobility of the glycinate ion is very much less than the chloride ion in the separation gel, so that a sharp boundary is maintained between these ions with the proteins of intermediate mobility sandwiched in between. This sharp boundary is maintained during electrophoresis and, at the pH of the separation gel (pH 8.9), the glycine mobility is greater than that of the protein, so the buffer boundary always runs ahead of the molecules being separated. Bromophenol blue is incorporated into one of the gels as a marker and this dye marks the boundary between the glycinate and chloride ions. The buffer system is therefore a *dis*continuous one and this gives rise to another name for this separation method, namely *disc electrophoresis*. This particular method of electrophoresis gives very sharp bands, as can be seen in the following experiment.

Toxicity   Acrylamide is toxic as marked on the reagent bottle and should be handled with care, particularly avoiding contact of the material with the skin.

MATERIALS <u>10</u>

Stock solutions   Prepare all the solutions except A and B in 100 ml volumetric flasks with distilled water, filter and store in brown bottles at 4°C

A.	Stock tris buffer: tris, 6 g; glycine, 28.8 g made up to 1 litre with water (pH 8.3)	1 litre
B.	Working tris buffer: dilute the stock solution 1 in 10 with distilled water	4 litres
C.	Tris 36.6 g; 1 mol/litre HCl, 48 ml; TEMED, 0.23 ml and water to 100 ml (pH 8.9)	100 ml
D.	Tris, 6.0 g; 1 mol/litre HCl, 48 ml; TEMED, 0.46 ml and water to 100 ml (pH 6.7, adjust with HCl if required)	100 ml
E.	Acrylamide, 28 g; bis, 0.74 g, water to 100 ml. *Care*: toxic!	100 ml
F.	Acrylamide, 10 g; bis, 2.5 g, water to 100 ml. *Care*: toxic!	100 ml
G.	Riboflavin, 4 mg, water to 100 ml	100 ml
H.	Sucrose (40 g/100 ml)	100 ml
J.	Ammonium persulphate (0.14 g/100 ml)	100 ml

Working solutions   Prepare the working solutions by mixing the stock solutions in the following proportions.

(i)	Small pore solution: 1 part of C, 2 parts of E and 1 part of water, pH 8.9	40 ml
(ii)	Large pore solution: 1 part of D, 2 parts of F, 1 part of G and 4 parts of H, pH 6.7	16 ml

Samples
Serum: albumin, 5 mg/ml; γ-globulin, 2 mg/ml; transferrin, 2 mg/ml   2 ml each

Staining solutions

Amido black, 1 g/litre in 7 per cent v/v acetic acid	500 ml
Coomassie blue, 2.5 g/litre in 200 g/litre TCA	500 ml
Acetic acid, 7 per cent v/v	3 litres
Tracking dye (bromophenol blue, 0.1 g/litre)	—

Equipment

Polyacrylamide-gel electrophoresis apparatus using rods of polyacrylamide	5
Fluorescent light	5
Syringe and needle	5

METHOD
Place rubber caps over the bottom of the hollow glass tubes and prepare the gels as given below.

Separation gel (7 per cent polyacrylamide)   Mix equal volumes of the catalyst solution of ammonium persulphate (J) and the small pore solution (i) and transfer 0.9 ml to each glass tube. Carefully add a water overlay and leave the gels to set (25–40 min).

Spacer gel (2.5 per cent polyacrylamide)   Remove the water overlay from the rods of polyacrylamide and layer 0.15 ml of the large pore solution over the top. Overlay the gel solution with water and place the tubes under a fluorescent light until gelation is complete (20–50 min). Finally cover the gel with a layer of dilute tris buffer, pH 8.3 (B).

Sample application   Mix the following samples with some of the 40 per cent w/v sucrose (H) so that they will layer on top of the gel underneath the buffer. Add some bromophenol blue to one of the serum samples to mark the ion boundary.

Tube no.	Sample
1 and 2	50 μl of serum
3 and 4	50 μl of albumin (5 mg/ml)
5 and 6	50 μl of γ-globulin (2 mg/ml)
7 and 8	50 μl of transferrin (2 mg/ml)

Electrophoresis   Remove the rubber caps at the bottom of the tubes, connect up the buffer reservoirs, and carry out electrophoresis at 5 mA/tube until the bromophenol blue has migrated almost to the end of the tube.

Staining   Switch off the current. Remove the gels by inserting a syringe needle between the gel and the wall of the tube and carefully discharge water from the syringe while rotating the tube. Collect the gels in test tubes and stain with amido black or Coomassie blue. Destain the gels by repeatedly washing in 7 per cent v/v acetic acid until the background is clear. Make a sketch of the separations and compare the results for serum with those obtained for cellulose acetate electrophoresis in Exp. 5.2.

Experiment 5.5   Isolation of lactate dehydrogenase isoenzymes by starch-block electrophoresis

PRINCIPLE

Isoenzymes are proteins that catalyse the same chemical reaction but which differ in a number of physico-chemical and kinetic properties. Lactate dehydrogenase has up to five isoenzymes which can be separated by electrophoresis. The advantage of starch-block electrophoresis is that the isoenzymes can be separated in sufficient quantities for their properties to be examined. A marker solution of bovine serum albumin

stained with bromophenol blue and human haemoglobin is included to check for
any drift of the pattern due to electro-osmosis.

MATERIALS   10
1. Extracts of rat tissues as in Exp. 12.8   —
2. Sodium phosphate buffer (0.1 mol/litre, pH 7.4)   6 litre + 5 ×
                                                                          tank
                                                                        capacity
3. Potato starch   3 kg
4. Perspex former for block (30 cm × 12 cm × 1 cm)   5
5. Absorbent gauze   —
6. Electrophoresis power pack   5
7. Marker solution of bovine serum albumin and human haemoglobin.   —
    (Dissolve bovine serum albumin in haemolysed erythrocytes and
    add bromophenol blue to stain the albumin)
8. Reagents for the spectrophotometric assay of lactate dehydrogenase   —
    (Exp. 12.3)
9. Sintered glass funnels   50

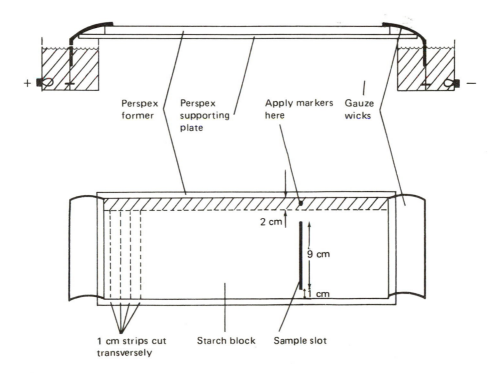

Fig. 5.4   Starch-block electrophoresis

The quantities are calculated assuming that each of the five pairs of students separates lactate dehydrogenase from one tissue.

METHOD

Wash 500 g of potato starch by decantation twice in distilled water and twice in the phosphate buffer. After the final wash, allow the starch to settle for about 2 h, remove the excess buffer by decantation, and compress the starch into a block using the Perspex former. Remove excess liquid from the block by blotting with Whatman 3 MM paper. Connect the block to the electrode compartment with several thicknesses of absorbent gauze saturated with buffer solution. Apply a potential of 6–8 V/cm to allow the system to equilibrate, then insert 1 ml of the tissue extract into a groove (9 cm) cut horizontally in the block 22 cm from the anodic end and starting 1 cm from the nearest edge. At the same time, apply the marker solution at the same distance from the anode and 1 cm from the furthest edge (Fig. 5.4). Fill the groove with starch and cover the block with a Perspex plate for safety.

Place the whole of the apparatus in the cold room or a refrigerator and apply a potential of 6–8 V/cm for 20 h. After electrophoresis, measure the distance migrated and the spread of the markers, then remove the part of the block containing the markers and discard it (Fig. 5.4). Cut the rest of the block transversely into strips 1 cm wide and transfer each strip to a sintered glass funnel. Pack the starch tightly with a flattened glass rod, elute twice with 2 ml of ice-cold phosphate buffer pH 7.4, and record the final volume of the eluate. Remove a convenient sample (0.1–1 ml) for the assay of the dehydrogenase activity and prepare a diagram of the total activity in each fraction against the distance from the cathode. Assay the crude extract and work out the recovery from the block.

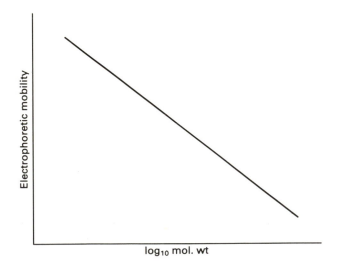

Fig. 5.5   The variation of mobility with the molecular weight of proteins on SDS-gels

If rat tissues are not readily available, then commercial ox heart and rabbit muscle lactate dehydrogenase can be used instead.

Experiment 5.6   SDS-gel electrophoresis

PRINCIPLE

Sodium dodecyl sulphate $(CH_3(CH_2)_{10}CH_2OSO_3^- Na^+)$ abbreviated SDS is a detergent that readily binds to proteins. At pH 7 in the presence of 1 per cent w/v SDS and 2-mercaptoethanol, proteins dissociate into their subunits and bind large quantities of the detergent. Under these conditions, most proteins bind about 1.4 g of SDS per gram of protein which completely masks the natural charge of the protein giving a constant charge to mass ratio. The larger the molecule therefore the greater the charge so the electrophoretic mobility of the complex depends on the size (mol. wt) of the protein and a plot of log mol. wt against relative mobility gives a straight line (Fig. 5.5). In this present experiment, the molecular weight of a protein is determined by comparing its mobility with a series of protein standards.

The sieving effect of the polyacrylamide is important in this technique and the range of molecular weights that can be separated on a particular gel depends on the 'pore size' of the gel. The amount of cross-linking and hence pore size in a gel can be varied by simply altering the amount of acrylamide to make 5 per cent or 10 per cent gels.

MATERIALS $\quad \frac{10}{5}$

1. PAGE apparatus for rods of polyacrylamide
2. SDS-buffer (sodium phosphate, 0.2 mol/litre, pH 7.1 containing 0.2 per cent w/v SDS) — 1 litre
3. TEMED (5 per cent v/v) — 25 ml
4. Acrylamide solution. (Dissolve 50 g of acrylamide and 1.36 g of bis in water to 200 ml. Filter if necessary and store in a dark bottle. *Caution!* the monomer is highly toxic, avoid inhalation and skin contact) — 200 ml
5. Sample buffer (sodium phosphate, 0.1 mol/litre, pH 7.1 containing 1 per cent w/v SDS and 1 per cent w/v 2-mercaptoethanol) — 50 ml
6. Ammonium persulphate (1 per cent w/v, prepare fresh) — 100 ml
7. Water baths at 37°C — 5
8. Vacuum flask (200 ml) — 5
9. 2-Mercaptoethanol — —
10. Bromophenol blue (0.25 per cent w/v in the sample buffer) — 100 ml
11. Glycerol — —
12. Coomassie blue (0.2 per cent w/v in methanol:acetic acid:water, 5:1:5). Filter and store in a dark bottle — 2 litres
13. Acetic acid (7 per cent w/v) — 2 litres

14. Standard proteins   —

Protein	Mol. wt	log_{10}mol. wt
Insulin	5 700	3.76
Cytochrome c	12 400	4.09
Myoglobin	17 200	4.24
Ovalbumin	45 000	4.65
Transferrin	77 000	4.89
Bovine serum albumin		
(monomer)	66 000	4.82
(dimer)	132 000	5.12
(tetramer)	264 000	5.42

15. α-Chymotrypsin   —

METHOD

Preparation of gel   The gels are prepared by mixing the components to give 5 per cent and 10 per cent polyacrylamide. The first three components are mixed together and deaerated in a vacuum flask for a few minutes. The ammonium sulphate and TEMED are then added and the mixture immediately introduced into the tubes. Distilled water is then added down the side of the tube to ensure a flat surface to the gel and to exclude air.

Component	5% gel (ml)	10% gel (ml)
Distilled water	10.4	0.4
SDS-buffer	25.0	25.0
Acrylamide solution	10.0	20.0
Ammonium persulphate	3.6	3.6
TEMED	1.0	1.0
Total volume	50.0	50.0

The gels polymerize in a few hours but are best if kept overnight and used the next day. In order to minimize evaporation and prevent contamination from atmospheric dust, a layer of parafilm is placed over the top of the gels.

Preparation of samples   Dissolve about 1 mg of the protein in 1 ml of the sample buffer and incubate for 3 h at 37°C or heat for 2 min at 100°C in a fume chamber. Some proteins are sensitive and require the milder treatment and a comparison can be made of the two methods of denaturation.

Electrophoresis   Remove the water from the surface of the gel and apply a sample of protein to the top of the gel. For the application, 100 µl of the protein solution is mixed with 45 µl of water, 45 µl of the SDS-buffer, 5 µl of 2-mercaptoethanol, 10 µl of bromophenol blue and 15 µl of glycerol. The 2-mercaptoethanol maintains a reducing environment, the bromophenol blue is present as a tracking dye and the glycerol increases the density of the mixture. The rest of the tubes are then filled with half strength SDS-buffer and electrophoresis carried out at 8 mA per tube until the tracking dye is almost at the end of the gel. The buffer used in the electrode compartments is prepared by diluting the SDS-buffer with an equal volume of distilled water.

Fixing and staining   Carefully remove the gels from the tubes with a fine hypodermic syringe and wash with distilled water. This procedure requires practice to avoid breaking the gel rods. Measure the length of each gel and the distance migrated by the tracking dye then immerse in a solution of Coomassie blue overnight.

Remove the dye solution, wash the gel with water and destain by washing repeatedly with 7 per cent w/v acetic acid.

Mobility determination   Measure the distance travelled by each protein and calculate the mobility relative to the tracking dye

$$\text{Mobility} = \frac{\text{Distance moved by protein}}{\text{Distance moved by bromophenol blue}}$$

Check for any shrinkage in the gels following the staining and allow for this in the calculation of the mobility.

Plot a curve of mobility against $\log_{10}$ mol. wt for the 5 per cent and the 10 per cent gels and determine the molecular weight of the chymotrypsin.

Experiment 5.7   Gradient-gel electrophoresis of proteins and multiple molecular forms of acetylcholinesterase

PRINCIPLE

If electrophoresis is carried out on a concave gradient of polyacrylamide, molecules will migrate through the slab until the pore size becomes too small for them to move any further. Electrophoresis is therefore carried out for a prolonged period, usually 24 h, so that effectively no futher migration takes place. Under these conditions, the distance migrated depends on the size and shape of the molecule and not on its charge. A plot of migration distance against $\log_{10}$ mol. wt gives a straight line if the molecules are approximately spherical and an example of an actual experiment is given in Fig. 5.6.

SDS is not used in this method of separation and so the molecular weights obtained are those of the native rather than the denatured protein subunits.

Multiple molecular forms of enzymes can therefore be separated by this technique and their position on the gel can be detected by a suitable stain. In the present experiment, acetylcholinesterase from rat brain is separated into a number of molecular weight species and the zones of activity are detected by incubating with acetylthiocholine and precipitating the released thiocholine as the copper salt.

$$2CH_3COSCH_2CH_2\overset{+}{N}(CH_3)_3 + (NH_2CH_2COO)_2Cu + H_2O$$

acetylcholinesterase

$$Cu[SCH_2CH_2\overset{+}{N}(CH_3)_3]_2 + 2NH_2CH_2COOH + 2CH_3COOH$$
Copper thiocholine
(white ppt)

MATERIALS   <u>10</u>
1. PAGE apparatus for use with slabs of polyacrylamide   5
2. Polyacrylamide slabs prepared as a concave gradient from 4 per cent to   5
   24 per cent polyacrylamide (available from Pharmacia and other
   commercial suppliers)
3. Extract of rat brain (Exp. 4.10a)   —
4. Electrophoresis buffer (88.7 mmol/litre tris, 81.5 mmol/litre boric acid, 5 litres
   2.5 mmol/litre EDTA, pH 8.3)
5. Staining solution for protein (Exp. 5.6)   —
6. Standard proteins (approx 1 mg/ml)   —

Protein	Mol. wt	log_{10}mol. wt
Bovine serum albumin		
(monomer)		
(dimer)	132 000	5.12
(tetramer)	264 000	5.42
Lactate dehydrogenase	136 000	5.13
Catalase	250 000	5.40
Apo-ferritin	450 000	5.65
Thyroglobulin	670 000	5.83

7. Staining solution for acetylcholinesterase. Dissolve 200 mg of   —
   acetylthiocholine iodide in 16 ml of water and add 28 ml of 0.1 mol/litre
   copper acetate solution dropwise with stirring. Centrifuge to remove
   the precipitate and dissolve 120 mg of glycine in the supernatant. Add
   2 mol/litre sodium acetate to give a final pH of 6.5–7.0

METHOD

Electrophoresis   Place the gels in the electrophoresis apparatus without the samples and connect to the power pack. Set the voltage on 100 V and leave until the current falls to a value between 35 and 40 mA. Apply the samples (30 μl) to the top of the gels

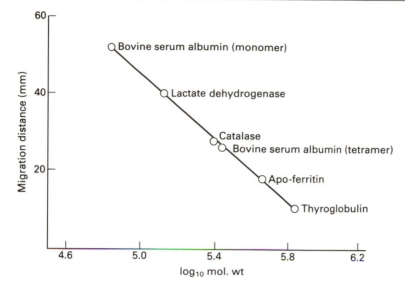

Fig. 5.6 Calibration curve for gradients of polyacrylamide-gels

in a plastic spacer arm which allows up to 14 samples to be run per gel. Overlay the samples with buffer and carry out electrophoresis at 100 V for 24 h at room temperature. Some means of cooling the apparatus should be employed and the easiest method is to circulate the electrophoresis buffer through a heat exchanger immersed in a bath of cold water.

Staining   Ideally the gel should be sliced vertically to produce two slabs of the same size but half the thickness but this may not always be possible and reasonable results can be obtained by staining the whole gel. Remove that part of the gel that contains the brain extract and immerse in the staining solution for acetylcholinesterase for 24 h at room temperature, then destain with 7 per cent w/v acetic acid. Stain the remainder of the gel for protein as previously described (Exp. 5.6). Plot the distance migrated against the $\log_{10}$ of the molecular weights of the proteins and comment on the range of species obtained for the acetylcholinesterase.

Further reading

Andrews, A. T., *Electrophoresis: Theory Techniques and Biochemical and Clinical Applications*, 2nd edn. Clarendon Press, Oxford, 1986.

Hames, B. and Rickwood, D., *Gel Electrophoresis of Proteins: A Practical Approach*. IRC Press, Oxford, 1981.

Smith, I., *Chromatographic and Electrophoretic Techniques*, 4th edn. Vol 2 *Electrophoresis*. Heinemann, London, 1976.

Davis, M. G. 'Electrophoretic Techniques' in *A Biologist's Guide to Principles and Techniques of Practical Biochemistry*, K. Wilson and K. H. Goulding (eds). 3rd edn. Arnold, London, 1986.

6. Immunochemical methods

Immunology

The immune response

Living organisms are able to defend themselves against invading foreign substances and the immune system is the means whereby they are able to recognize and reject foreign cells and their products. The study of this immune response is the branch of biological science known as immunology and this can be divided into two broad subdivisions, namely cellular immunity and humoral immunity.

Cellular immunity This involves the ability of the cells to recognize alien substances and to respond by ingesting them so that they are effectively removed from the organism.

Humoral immunity This is concerned with complex proteins in the blood plasma which are able to react with and neutralize soluble foreign compounds of a high molecular weight. It is this aspect of immunology that we shall be concerned with in this chapter.

The antibody–antigen reaction

Antibody The plasma proteins produced by lymphocytes in response to the presence of external or non-self molecules are known as immunoglobulins or, more generally, antibodies.

Antigens These are foreign compounds that provoke the formation of antibodies. The chemical nature of antigens is quite diverse and they are usually proteins or polysaccharides although lipoproteins and lipopolysaccharides are also known. These macromolecules may occur free in solution or they may be bound to the surface of cells or particles such as viruses, bacteria, pollens or tissues such as erythrocytes or kidney from another individual.

Antigens are usually macromolecules although low molecular weight compounds such as drugs can become antigenic when bound to a protein. These small molecules which in themselves are not antigenic are known as *haptens*.

Specificity An antibody (Ab) produced in response to a foreign substance reacts with the antigen (Ag) to form an immune complex (Ab–Ag) and is thereby inactivated. In most cases, the immune complex is insoluble and is removed from the circulation.

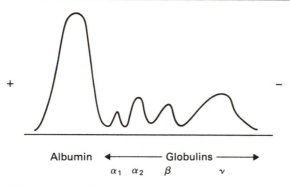

$+$ $-$

Albumin ←——— Globulins ——→

α_1 α_2 β γ

Fig. 6.1 Scan of a stained cellulose acetate electrophoresis strip of human serum proteins

$$Ab + Ag \longrightarrow (Ab-Ag)_n$$

This reaction is highly specific and the antibody is bound by close range non-covalent Van der Waals' forces to a small site on the antigen known as the *antigenic determinant*. A single antigenic molecule may contain a number of different antigenic determinants and bovine serum albumin (BSA) has several hundred antigenic determinants per molecule.

Antibodies are also specific in that they normally react only with foreign substances and not with the plasma and tissues of the individual. This ability to recognize 'self' as opposed to 'non-self' is extremely important and problems occur when antibodies are made against one's own tissue giving rise to *autoimmune diseases* such as rheumatoid arthritis and ulcerative colitis.

Immunoglobulins

Classification Chemically antibodies are proteins that are globulins with an immunological activity, hence the name immunoglobulins. Electrophoresis of blood serum on cellulose acetate gives rise to a number of protein bands (Exp. 5.2) and if the strip is stained and then scanned, a pattern similar to that of Fig. 6.1 is obtained. The immunoglobulins are all found in the γ-globulin region but this band is quite diffuse and includes several different types of antibody each with a different structure and function (Table 6.1).

Table 6.1 The classification of the human immunoglobulins

Class	Serum mg/ml	Mol. wt	Light chain	Heavy chain	Structure*
IgG	12	150 000	κ or λ	γ	$\kappa_2\gamma_2$
IgA	3	$>300\,000$	κ or λ	α	$(\kappa_2\alpha_2)_n$
IgM	1	950 000	κ or λ	μ	$(\kappa_2\mu_2)_5$
IgD	0.1	160 000	κ or λ	δ	$\kappa_2\delta_2$
IgE	0.001	200 000	κ or λ	ε	$\kappa_2\varepsilon_2$

* Alternative structures are possible with the κ chains replaced by λ chains.

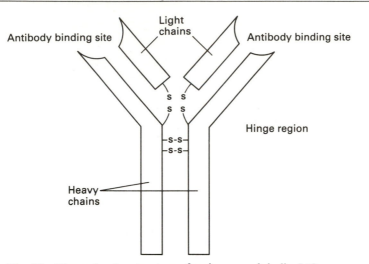

Fig. 6.2   The molecular structure of an immunoglobulin, IgG

Structure of IgG   The highest serum concentration of the immunoglobulins is that of IgG and this is a Y-shaped molecule made up of two light (mol. wt 23 000) and two heavy (mol. wt 52 000) chains linked by disulphide bonds (Fig. 6.2). The amino acid sequence at the C-terminal end of the chain is constant while the N-terminal region is highly variable and it is this variability that accounts for the immense diversity of antibodies. The antibody-combining sites are located at the end of the heavy and light chains and the hinge region gives flexibility to the molecule so that sites which are different distances apart can be bound.

The structures of the other immunoglobulins are similar to that of IgG with some variation.

Immunochemical assays

Functional assays   Biological fluids contain a complex mixture of proteins and in order to assay a particular protein, it is necessary to use a system that reacts with that protein and only that protein. The use of antibodies raised against specific proteins means that such assays are possible so that immunochemical techniques are accurate and highly specific.

There are a number of assay systems currently in use but they all depend on the formation of an immune complex between the antibody and the antigen. The reaction is reversible and a typical 'titration curve' of increasing amount of antigen added to a fixed concentration of antibody is shown in Fig. 6.3. It is important to be aware of the shape of this curve and to identify clearly the antigen excess region if results are not to be misinterpreted.

Assays available   Immunoassays involve the measurement of one or other of the

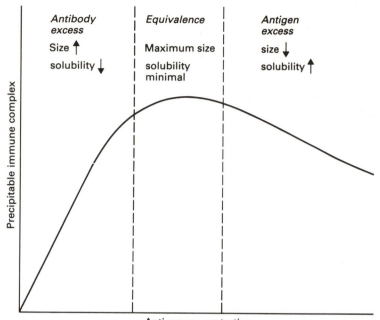

Fig. 6.3   A typical precipitation curve for an antiserum titrated against a fixed concentration of antibody

components of the antibody–antigen system, namely, antibody, antigen and the immune complex. There are a very large number of assays possible with a given combination of antibody and antigen and some examples of those involving the quantitation of the immune complex are given below:

1. *Recovery of immune complex*
    (a) Assay by weight
    (b) Assay of volume (immunocrit)
    (c) Measurement of protein or nitrogen
2. *Gel precipitation*
    (a) Radial immunodiffusion (Mancini)
    (b) Double diffusion (Ouchterlony)
    (c) Immunoelectrophoresis
3. *Labelled antigen or antibody*
    (a) Radioimmunoassay
    (b) Haemagglutination
    (c) Enzyme linked immunoabsorbent assay (ELISA)
4. *Light scattering*
    (a) Immunonephelometry
    (b) Immunoturbidimetry

5. *Other*
   (a) Changes in viscosity
   (b) Changes in sedimentation

The major differences between the assays are the sensitivities and the time taken to carry them out. As expected, the most sensitive methods are usually the most difficult technically and therefore the ones that can give the greatest error.

Immunochemical assays are now used extensively in the biological sciences and some examples of those currently in use are given in the following series of experiments.

Human serum

Human serum is required for most of the experiments and one person in each pair should volunteer to give about 10 ml of blood. The collection of the blood must be carried out by a qualified member of staff and the usual precautions taken when handling body fluids (Chapter 1).

Collect the blood into a glass, not plastic, container and leave the blood to clot at 37°C for 1 h. Cool on ice to allow the clot to retract and separate the serum by decanting the liquid from the clot and centrifuging on a bench centrifuge. Store the serum at 0°C until required then warm to room temperature just before use.

Experiments

Experiment 6.1   Quantitative precipitin test

PRINCIPLE

The γ-globulin fraction is isolated from human serum by salt precipitation and used as the antigen in the experiment. Increasing quantities of antigen are mixed with a constant amount of antiserum and the protein precipitated in the immune complex assayed by measuring the extinction at 280 nm.

Maximum interaction between the antibody and the antigen is only obtained if the incubation at 4°C is carried out for several days but with class experiments this is not very convenient and quite reasonable results can be obtained after 1h.

MATERIALS	100
1. Fresh human serum	100 ml
2. Saturated ammonium sulphate (Add 155 g of ammonium sulphate to 200 ml of water)	200 ml
3. UV spectrophotometers	10
4. Buffered saline. (Dissolve 8 g NaCl, 0.2 g KCl, 1.5 g Na_2HPO_4 and 0.2 g KH_2PO_4 in water. Make up to 1 litre then autoclave for 20 min)	1 litre
5. NaOH (0.1 mol/litre)	2 litres
6. Antiserum (rabbit anti-human whole serum)	—

METHOD

Pipette 2 ml of human serum into a test tube and slowly add 2 ml of saturated ammonium sulphate solution; leave on ice for about 30 min with occasional stirring then centrifuge at 2500 g for 15 min, remove the supernatant and resuspend the pellet in a known volume of the buffered saline. Measure the extinction at 280 nm on a suitably diluted sample and calculate the amount of protein present assuming that 1 mg/ml has an extinction of 1.40 at 280 nm.

Add the following μg of 'antigen' to a series of tubes: 0, 10, 20, 50, 100, 200, 300, 400, 500, 1000 then buffered saline to bring the final volume to 0.45 ml. Add 50 μl of antiserum to each tube, mix thoroughly then incubate at 37°C for 1 h, then 4°C for 1 h. Centrifuge the tubes at 2500 g for 5 min and remove the supernatant. Wash the precipitate twice with ice-cold buffered saline and dissolve the final pellet in 0.1 mol/litre NaOH.

Read the extinction of each tube at 280 nm and plot a graph of the extinction against the amount of antigen present. Finally, determine the antibody content per ml of antiserum.

Experiment 6.2   Ouchterlony immunodiffusion

PRINCIPLE

Human and animal sera often contain antibodies which will react with antigens from other species of animals because of exposure to the cross-reacting antigens during microbial infections. One example of this is the agglutinins against blood group substances which are thought to be acquired during infection by enteric bacteria

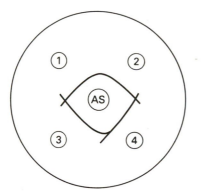

Fig. 6.4   Patterns of precipitation obtained with Ouchterlony immunodiffusion

AS is antiserum placed in the centre well. The test serum samples are placed in the surrounding wells (1, 2, 3, 4).
Fused precipitation lines: Serum samples 1 and 2 contain some common antibodies.
Crossed precipitation lines: Serum samples 1 and 3 and 2 and 4 do not have common antibodies.
Spur plus fused precipitation lines: Serum samples 3 and 4 have some common antibodies but 4 has a unique antibody not shared with 3.

whose surface antigens cross-react with antigens A and B of the human blood group substances.

In this experiment, rabbit antiserum is examined for the presence of cross-reacting precipitins against human, rabbit and rat serum by immunodiffusion.

Immunodiffusion involves layering a glass plate with agar-gel and cutting wells into the gel to form a pattern of the type shown in Fig. 6.4. The actual number of wells and their size and distance apart depends on the antigens and antisera being examined. The antiserum or antibody is placed in the centre well and the test compounds in the surrounding wells. During incubation, the antibody and antigen react after diffusing through the gel and form a precipitate where they meet. The immune complex is soluble in excess antigen (Fig. 6.3) and so the equivalence point is marked by a sharp line. The position of this line depends on the relative concentration of antibody and antigen in the agar which in turn depends on the concentration in the well, the molecular weights and the diffusion coefficients of the constituents. If the antibody and antigen contain several reactive species then multiple lines of precipitation will be seen.

MATERIALS	100
1. Saline (0.9 per cent w/v NaCl; 0.1 per cent w/v Na azide)	12 litres
2. Agar (1.5 per cent w/v in saline; heat the solution to 90°C until all the clumps of agar have dissolved. Cool to 55°C then pour into plastic Petri dishes to a depth of 6 mm). Allow for four dishes per pair	—
3. Human serum	50 ml
4. Rabbit serum	50 ml
5. Rat serum	50 ml
6. Rabbit antihuman serum	50 ml
7. Incubators at 37°C	10
8. Gel punch pattern (Fig. 6.4)	4

METHOD

Allow the plates to cool then cut out six wells round a central well with a template. Remove the gel plugs by gentle suction and place your own serum in the centre well and the animal and other human sera in the surrounding wells until they are just full. Set up another plate with different dilutions of the sera then cover the Petri dishes and incubate at 37°C in a moist chamber overnight.

Examine the patterns obtained and interpret the results.

Experiment 6.3   Purification of immunoglobulins by ion exchange chromatography

PRINCIPLE

When human serum is added to an ion exchange column such as DEAE–Sephadex, most of the serum proteins become bound to the column. However, by using a buffer with a pH greater than 6.5, IgG is not absorbed and is eluted in the first fractions. The other proteins are then removed from the column by a stepwise elution with increasing concentrations of salt. The isolated fractions are then analysed by cellulose acetate electrophoresis (Exp. 5.2) and immunoelectrophoresis (Exp. 6.4).

MATERIALS | | 10
1. DEAE–Sephadex A50–120 | 2 g
2. Disposable syringes (10 ml) | 5
3. Tris–HCl buffer (10 mmol/litre, pH 8.0) | 3 litres
4. Buchner filtration apparatus | 5
5. NaOH (1 mol/litre) | 1 litre
6. HCl (1 mol/litre) | 1 litre
7. Glass wool or plastic sinters | —
8. Peristaltic pumps that can pump less than 60 ml/h | 5
9. Human serum | 5 ml
10. UV spectrophotometers or column monitoring equipment | 5
11. Buffered saline solutions. (Prepare the following concentrations of saline in the tris–HCl buffer: | 250 ml
    (a) 20 mmol/litre
    (b) 50 mmol/litre
    (c) 100 mmol/litre
    (d) 150 mmol/litre
    (e) 200 mmol/litre
    (f) 300 mmol/litre
12. Solid polyethyleneglycol | 500 g
13. Materials for Exp. 5.2 | —
14. Materials for Exp. 6.4 | —

METHOD

Preparation of column   Add 2 g of DEAE–Sephadex A50–120 to 70 ml of water and leave until swollen. Degas the suspension under vacuum and leave to equilibrate to room temperature. Place some glass wool in the bottom of a 10 ml disposable syringe as a support and prepare a column of the ion exchange material. Wash the column with 1 mol/litre NaOH then repeatedly with water until the pH falls to about 7. Wash with saline then 1 mol/litre HCl and finally with the tris–HCl buffer until the eluate registers pH 8.0. Alternatively the material can be washed on a Buchner funnel.

Separation of immunoglobulins   Load 0.5–1.0 ml of your own serum on to the DEAE–Sephadex and collect five fractions of about 2.5 ml each by pumping the tris–HCl buffer through the column. In the meantime prepare 25 ml of a series of fluids containing sodium chloride at the following concentrations in the buffer: 20 mmol/litre,   50 mmol/litre,   100 mmol/litre,   150 mmol/litre,   200 mmol/litre, 300 mmol/litre. Pump these solutions through the column in turn starting with the lowest salt concentration and collect fractions until all the protein has been eluted with that solution, then move on to a higher concentration of salt. Carefully monitor the proteins eluted by measuring the absorbance of each fraction at 280 nm (Exp. 7.9) and plot the extinction against fraction number.

Concentrate each fraction between 5 and 10 times by dialysing against solid

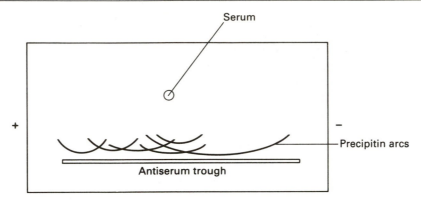

Fig. 6.5   Immunoelectrophoresis of serum

polyethyleneglycol for about 3 h and examine the protein concentrates by cellulose acetate electrophoresis, immunodiffusion and immunoelectrophoresis.

Experiment 6.4   Immunoelectrophoresis

PRINCIPLE

Immunoelectrophoresis consists of two methods that are combined, namely agar-gel electrophoresis and immunodiffusion. The agar plate has a longitudinal trench cut which accommodates the antiserum after electrophoresis and a central well or wells to accommodate the test material (Fig. 6.5). The serum sample is placed in the well and electrophoresis carried out for 4–6 h at a field strength of 3–6 V/cm. The protein components separate according to their charge (Chapter 5) but unlike normal electrophoresis, the bands are not stained for protein. Instead antiserum is pipetted into the longitudinal trough and the plate incubated in a humid chamber. Diffusion of the antiserum and protein occurs and, where they meet, precipitin arcs are formed that give rise to a characteristic pattern.

MATERIALS	10
1. Barbiturate buffer (0.07 mol/litre, pH 8.6)	1 litre
2. Agar solution (1 per cent w/v in the barbiturate buffer: heat the agar at 90°C until all the lumps are dissolved and pour at 55°C as described under Method)	—
3. Glass slides (7.5 cm × 5 cm)	30
4. Incubator at 37°C	1
5. Spirit level	1
6. Moist storage cabinets	2
7. Shandon cutting device	2
8. Tracking dye (bromophenol blue, 0.1 g/litre)	—
9. Electrophoresis equipment	5
10. Saline (0.9 per cent w/v NaCl; 0.1 per cent w/w Na azide)	100 ml

11. Human serum   —
12. Rabbit anti-human serum   —
13. Protein fractions from Exp. 6.3   —

METHOD

Each pair should heat six glass plates in an oven at 100°C then remove them and place them on a level surface. Carefully pipette 7 ml of the hot agar solution on to the surface of the slides and with care the molten agar will set in a uniform layer and the surface tension will hold the agar at the edges.

Leave the plates to solidify for 15 min, then transfer them to a moist storage cabinet. Cut the wells and the troughs just before the plates are used with the Shandon cutting device to give troughs 5 cm long and 1 mm wide. Punch out the antigen wells in the agar between each pair of troughs with a 1 mm diameter needle. Attach the needle to a water pump vacuum and remove the troughs by gentle suction making sure that the sides are not damaged. Carry out immunoelectrophoresis on your own serum and the concentrated material from the column fractions in two runs in duplicate, one using neat material and the other using a one-fifth dilution of the serum or fraction. Place a drop of the tracking dye to mark the position of the albumin. Dilute the samples with saline containing azide and make sure that the sides of the wells are not damaged.

Apply a potential difference to give 3–6 V/cm and stop the electrophoresis when the bromophenol blue has run a distance of 1.5 cm. Remove the agar from the trough and add 50 µl of rabbit anti-human serum. Place the plate in a moist chamber for 24 h to allow immunodiffusion to occur and precipitin lines should be visible after 24 h.

Experiment 6.5   The determination of albumin by radial immunodiffusion

PRINCIPLE

Radial immunodiffusion, also known as the Mancini technique, is similar to that described in Exp. 6.2 but in this method the antiserum is incorporated into the gel. The

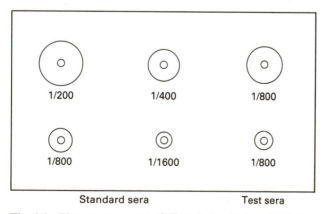

Fig. 6.6   The measurement of albumin by single radial immunodiffusion

antigen is placed in wells cut into the gel and diffuses out into the agar to form a circular immunoprecipitate (Fig. 6.6). Diffusion takes from 2 to 7 days depending on the molecular weight of the antigen and, when this is complete, the area of the circle is directly proportional to the concentration of the antigen. A standard curve is therefore constructed by plotting the square of the diameter of the ring against concentration and this can be used to determine the concentration of an 'unknown'.

The time taken for complete diffusion may be inconvenient for a class experiment and a 24 h incubation which gives a curve can be used (Fig. 6.7) to assay all but the highest concentration of antigen. Alternatively, the diameter of the ring can be plotted against the log of the antigen concentration following incubation for 4 h although the assay will inevitably be less accurate.

MATERIALS   10
1. Materials for immunodiffusion (Exp. 6.2)   —
2. Rabbit anti-human albumin   1 ml
3. Human serum   1 ml
4. Unknown serum albumins   1 ml
5. Tannic acid (1 per cent w/v)   100 ml

METHOD
Prepare the agar as described in Exp. 6.2 and allow to cool to about 55°C. Carefully add 0.4 µl of rabbit anti-human albumin per cm² of the plate area and pour on to the glass slides. Leave the plates undisturbed until the gel has set, then cut out the required number of wells (say 16) making sure they are not too near the edge of the plate.

Prepare dilutions of the standards (1/200, 1/400, 1/800, 1/1600) and tests (1/800) in saline and pipette 3 µl into the wells on the plate: leave on the bench for 30 min then

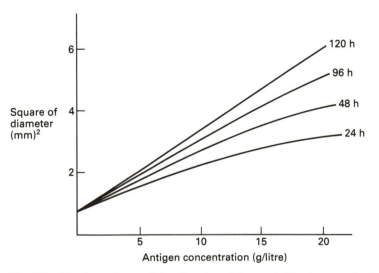

Fig. 6.7   The dependence of the diameter of the precipitin rings on the time of incubation

place in a moist chamber at 4°C for 24 h. At the end of this period, examine the slides and plot the square of the diameter of the rings against the concentration of antigen and use the standard curve to estimate the concentrations of albumin in the test samples.

The diameter of the rings can be measured directly by placing the slides on a black background and using side illumination. If the rings are not clear then the slides can be treated with 1 per cent w/v tannic acid to aid visualization.

Replace the plates in the moist chamber and measure the circles at 2, 4 and 7 days after the experiment. Again plot the square of the diameter of the rings on each occasion against the antigen concentration. What effect does the time of incubation have on the standard curve and why?

Experiment 6.6   The determination of albumin by Laurell rocket immunoelectrophoresis

PRINCIPLE

Rocket immunoelectrophoresis is very similar to radial immunodiffusion in that the antigen is incorporated into the agar-gel. This time, however, the samples move through the gel by electrophoresis rather than by simple diffusion.

The antiserum is incorporated into the agar and holes are cut along the origin to hold the test material and dilutions of a standard antigen for calibration. When a

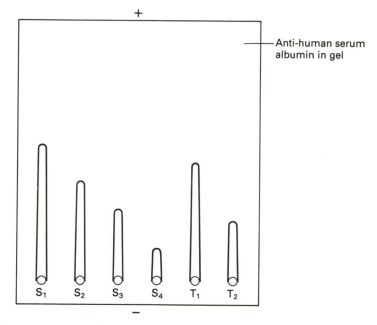

Standards: S_1, S_2, S_3, S_4
Tests: T_1, T_2

Fig. 6.8 Laurell rocket immunoelectrophoresis

potential is applied across the gel, the antigen migrates towards the anode while the antibody in the gel moves towards the cathode and an immune complex forms where they meet. Initially there is an excess of antigen so that the immune complex is soluble (Fig. 6.3) but as the experiment proceeds the antigen continues to migrate and becomes more dilute as more antigen is present in the immune complex. Later on equivalence is reached and an insoluble immune complex is formed. The complex itself migrates but eventually becomes stationary when all the antigen is present in the immune complex. The precipitate is rocket shaped and, for a given antibody concentration, there is a linear relationship between the distance moved by the precipitate and the concentration of the antigen (Fig. 6.8).

MATERIALS   <u>10</u>
1. Materials for immunoelectrophoresis (Exp. 6.4)   —
2. Materials for Exp. 6.5   —
3. Coomassie blue (0.25 per cent w/v in methanol:acetic acid:water,   1 litre
   5:1:5)
4. Destaining solution (methanol:acetic acid:water, 5:1:5)   1 litre
5. Hair driers   2

METHOD
Prepare the plates as for the immunoelectrophoresis but add 0.6 μl of rabbit anti-human albumin per cm^2 of the plate area. Cut 12 wells into the gel in a straight line 1.5 cm from one edge of the plate and place in the electrophoresis apparatus. Apply the diluted standards to four of the wells and 1/400 dilutions of the tests to the other wells. All dilutions are made in saline as in the previous experiment. Pipette 3 μl of the samples into each well as quickly as possible so as to minimize the amount of radial diffusion. When all the samples have been added to the plate, which should be set up in duplicate, place the lid of the tank on the apparatus and turn on the power. Carry out electrophoresis at 5–10 V/cm with the cooling system activated. At the end of this time, turn off the power and remove the plate. Place the slides under a number of filter papers with a light weight on top, leave for 10 min then remove the filter papers. Dry the plate completely in a stream of hot air from a hair drier then immerse the dry plate in Coomassie blue for 10 min. Remove the excess stain with the destaining solution and measure the heights of the rockets from the anodal edge of the well to the peak of the rocket. Plot the rocket height against albumin concentration to produce a standard curve, then use this curve to estimate the albumin concentrations in the eight serum samples.

Finally, compare the values obtained with those from the single radial immunodiffusion (Exp. 6.5) and comment on the two methods.

Experiment 6.7   The detection and assay of a myeloma protein

PRINCIPLE
Multiple myeloma is a malignant disorder of the bone marrow where a single cell multiplies to produce a very large number of cells of one kind. The resultant colony of

cells is a *clone* as they are all descended from one cell. The cells therefore have identical properties and produce large quantities of homogeneous immunoglobulin. The detection and quantitation of myeloma protein is important in the diagnosis of this condition and the following experiment shows how electrophoretic and immunological techniques are used in the clinical biochemistry laboratory.

Samples of serum from myeloma patients can be obtained by arrangement with the medical staff of a local chemical pathology laboratory and must be treated even more carefully than usual when handling body fluids. The actual type of myeloma will of course depend on what samples are available at the time.

MATERIALS   10
1. Sera from myeloma patients   —
2. Materials for cellulose acetate electrophoresis (Exp. 5.2)   —
3. Materials for immunoelectrophoresis (Exp. 6.4)   —
4. Anti-IgA serum   —
5. Anti-IgG serum   —
6. Materials for single radial immunodiffusion (Exp. 6.5)   —

METHOD

Detection   Use cellulose acetate electrophoresis (Exp. 5.2) to detect myeloma protein in the samples provided. Stain the cellulose acetate strips and compare the pattern obtained with normal human serum.

Demonstration   The class of the immunoglobulin which constitutes the myeloma protein can be demonstrated by immunoelectrophoresis in agarose-gel (Exp. 6.4). Use the anti-IgA and the anti-IgG sera to identify the immunoglobulin.

Quantitation   Single radial immunodiffusion (Exp. 6.5) is a technique that is frequently used to measure the amount of immunoglobulins present in the serum. Use two plates in duplicate for the experiment with anti-IgA in the gel of one pair and anti-IgG in the other pair.

Further reading

Haeney, M., *Introduction to Clinical Immunology*. Butterworths, 1985.
Hudson, L. and Hay, P. C., *Practical Immunology*. Blackwell Scientific Publications, Oxford, 1980.
McConnell, I., Munro, A. and Waldman, H., *The Immune System: A Course on the Molecular and Cellular Basis of Immunity*, 2nd edn. Blackwell Scientific Publications, Oxford, 1981.
Playfair, J. H. C., *Immunology at a Glance*, 3rd edn. Churchill Livingstone, Edinburgh, 1985.
Roitt, I. M., *Essential Immunology*. Blackwell Scientific Publications, Oxford, 1981.

7. Spectrophotometry

Colorimetry

Why solutions are coloured

Many biochemical experiments involve the measurement of a compound or group of compounds present in a complex mixture. Probably the most widely used method for determining the concentration of biochemical compounds is colorimetry, which makes use of the property that when white light passes through a coloured solution, some wavelengths are absorbed more than others (Fig. 7.1). Many compounds are not themselves coloured, but can be made to absorb light in the visible region by reaction with suitable reagents. These reactions are often fairly specific and in most cases very sensitive, so that quantities of material in the region of millimole per litre concentrations can be measured. The big advantage is that complete isolation of the compound is not necessary and the constituents of a complex mixture such as blood can be determined after little treatment. As discussed below, the depth of colour is proportional to the concentration of the compound being measured, while the amount of light absorbed is proportional to the intensity of the colour and hence to the concentration.

The Beer–Lambert law

When a ray of monochromatic light of initial intensity I_0 passes through a solution in a transparent vessel, some of the light is absorbed so that the intensity of the transmitted light I is less than I_0. There is some loss of light intensity from scattering by particles in the solution and reflection at the interfaces, but mainly from absorption by the solution (Fig. 7.2). The relationship between I and I_0 depends on the path

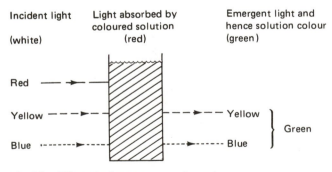

Fig. 7.1 Why solutions appear coloured

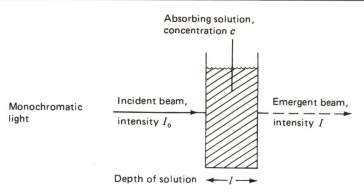

Lambert's law: $I = I_0 e^{-k_1 l}$

Beer's law: $I = I_0 e^{-k_2 c}$

Fig. 7.2   The absorption of light by a solution

length of the absorbing medium, l, and the concentration of the absorbing solution, c. These factors are related in the laws of Lambert and Beer (Fig. 7.2).

Lambert's law   When a ray of monochromatic light passes through an absorbing medium its intensity decreases exponentially as the length of the absorbing medium increases.

Beer's law   When a ray of monochromatic light passes through an absorbing medium its intensity decreases exponentially as the concentration of the absorbing medium increases.
   These two laws are combined together in the *Beer–Lambert* law:

$$I = I_0 e^{-k_3 cl}$$

Transmittance   The ratio of intensities is known as the transmittance (T) and this is usually expressed as a percentage.

$$\text{Per cent } T = I/I_0 \times 100 = e^{-k_3 cl}$$

This is not very convenient since a plot of per cent transmittance against concentration gives a negative exponential curve (Fig. 7.3(a)).

Extinction   If logarithms are taken of the equation instead of a ratio then:

$$\log_e I_0/I = k_3 cl$$

$$\log_{10} I_0/I = k_3 cl/2.303$$

or

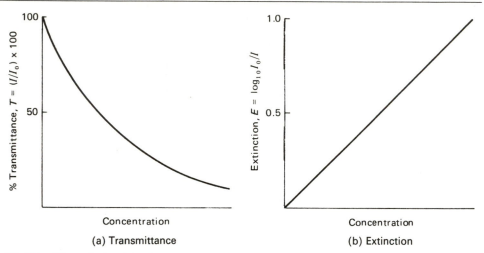

(a) Transmittance

(b) Extinction

Fig. 7.3 The relationship between the absorption of light and the concentration of an absorbing solution

$$\log_{10} I_0/I = kcl$$

The expression $\log_{10} I_0/I$ is known as the *extinction* (E) or *absorbance* (A). The extinction is sometimes referred to as the *optical density*, but this name is no longer recommended. Therefore,

$$E = kcl$$

If the Beer–Lambert law is obeyed and l is kept constant, then a plot of extinction against concentration gives a straight line passing through the origin (Fig. 7.3(b)) which is far more convenient than the curve for transmittance.

Some colorimeters and spectrophotometers have two scales, a linear one of per cent transmittance and a logarithmic one of extinction. It is this latter scale that is related linearly to concentration and is the one used in the construction of a standard curve (Fig. 7.4). With the aid of such a standard curve the concentration of an unknown solution can easily be determined from its extinction.

Linear scale of % transmittance

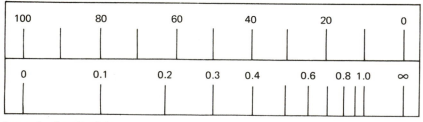

Log scale of extinction

Fig. 7.4 The relationship between per cent transmittance and extinction

Molar extinction coefficient   If l is 1 cm and c is 1 mol/litre then the absorbance is equal to k, the *molar extinction coefficient*, which is characteristic for a compound. The molar extinction coefficient k is thus the extinction given by 1 mol/litre in a light path of 1 cm and is usually written $E_{1\,cm}^{1\,mol/l.}$, it has the dimension of litre $mol^{-1}\,cm^{-1}$.

Specific extinction coefficient   The molecular weights of some compounds such as proteins or nucleic acids in a mixture are not readily available and, in this case, the *specific extinction coefficient* is used. This is the extinction of 10 g/litre (formerly known as 1 per cent w/v) of the compound in the light path of 1 cm, $E_{1\,cm}^{10\,g/l}$.

Limitations of the Beer–Lambert law   Sometimes, a non-linear plot is obtained of extinction against concentration and this is probably due to one or other of the following conditions not being fulfilled.

1. Light must be of a narrow wavelength range and preferably monochromatic.
2. The wavelength of light used should be at the absorption maximum of the solution: this also gives the greatest sensitivity.
3. There must be no ionization, association, dissociation, or solvation of the solute with concentration or time.
4. The solution is too concentrated, giving an intense colour. The law only holds up to a threshold maximum concentration for a given substance.

Measurement of extinction

The earliest colorimeters relied on the human eye to match the colour of a solution with that of one of a series of coloured discs. The results obtained were too subjective and not particularly accurate. Visual colorimeters are now of historical interest only and are not described here. The photoelectric cell is superior to the human eye in assessing the degree of absorption of a colour and is more objective.

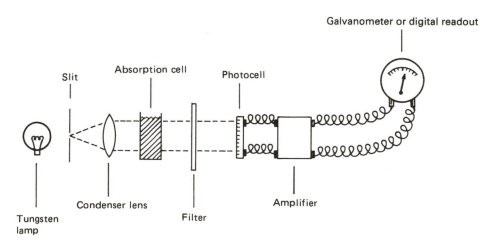

Fig. 7.5   A diagram of a photoelectric colorimeter

Table 7.1   The relationship between the colour of the solution examined and the filter chosen for colorimetric analysis

Colour of solution	Filter
Red–orange	Blue–blue green
Blue	Red
Green	Red
Purple	Green
Yellow	Violet

The photoelectric colorimeter   A diagram of the basic arrangement of a typical colorimeter is given in Fig. 7.5. White light from a *tungsten lamp* passes through a *slit*, then a *condenser lens*, to give a parallel beam which falls on the solution under investigation contained in an *absorption cell* or *cuvette*. The cell is made of glass with the sides facing the beam cut parallel to each other. In most cases, the cells are 1 cm square and will hold 3 ml of liquid comfortably.

Beyond the absorption cell is the *filter*, which is selected to allow maximum transmission of the colour absorbed. If a blue solution is under examination, then red is absorbed and a red filter is selected. The colour of the filter is, therefore, complementary to the colour of the solution under investigation (Table 7.1). In some instruments the filter is located before the absorption cell. The filters give narrow transmission bands and, therefore, approximate to monochromatic light.

The filter is chosen so that Beer's law is obeyed.

The light then falls on to a *photocell* which generates an electrical current in direct proportion to the intensity of light falling on it. This small electrical signal is increased in strength by the *amplifier*, and the amplified signal passes to a *galvanometer*, or *digital readout*, which is calibrated with a logarithmic scale so as to give absorbance reading directly. Photocells used in such cheap colorimeters have a precision of about 0.5 per cent. The blank solution is first put in the colorimeter and the reading adjusted to zero extinction; this is followed by the test solution and the extinction is read off directly.

A better method is to split the light beam, pass one part through the sample and the other through the blank, and balance the two circuits to give zero. The extinction is determined from the potentiometer reading which balances the circuit.

UV and visible spectrophotometry

Absorptiometric analysis

A spectrophotometer is a sophisticated type of colorimeter where monochromatic light is provided by a grating or prism. The band width of the light passed by a filter is quite broad, so that it may be difficult to distinguish between two compounds of closely related absorption with a colorimeter. A spectrophotometer is then needed, when the two peaks can be selected on the monochromator.

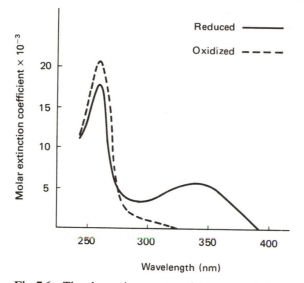

Fig. 7.6   The absorption spectra of the oxidized and reduced forms of NAD and NADP

Some compounds absorb strongly in the ultraviolet region and their concentration can be determined as with a colorimeter by using a more expensive type of spectrophotometer which operates down to 190 nm. For example, the concentration of uric acid can be estimated by measuring the extinction of the solution at 293 nm before and after treatment with an excess of the enzyme uricase. At pH 9 uric acid, which absorbs at 293 nm, is oxidized by uricase to allantoin, which has no absorption at this wavelength.

The most frequently used wavelength in the untraviolet region is probably 340 nm. At this wavelength, the reduced forms of the pyridine nucleotide coenzymes $NADH_2$ and $NADPH_2$ absorb strongly while the oxidized forms do not (Fig. 7.6). NAD has a typical dinucleotide structure:

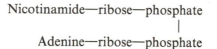

NADP has the same structure except for a phosphate on the 2′ position of the

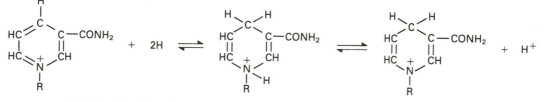

Fig. 7.7   The reduction of nicotinamide adenine dinucleotide

adenine ribose. The two H react with the nicotinamide part of the molecule but, at physiological pH, one H dissociates as shown (Fig. 7.7).

The reduced form thus has a quinonoid structure and it is this that is responsible for the absorption at 340 nm.

The progress of any enzyme-catalysed reaction involving these coenzymes can easily be observed by following the rate of appearance or disappearance of $NADH_2$ from the absorption at 340 nm. The method is very sensitive since the molar extinction coefficient of $NADH_2$ at 340 nm is 6.3×10^3 litres $mol^{-1} cm^{-1}$. This means that the conversion of 1 µmol substrate/ml is indicated by a change in absorbance of 6.3. Enzyme-catalysed reactions involving changes in only nanomoles of substrate per minute can therefore be readily followed.

Absorption spectra

Many compounds have characteristic absorption spectra in the ultraviolet and visible regions so that identification of these materials in a mixture is possible. This point will be illustrated many times throughout this book.

Proteins   Proteins absorb strongly at 280 nm according to their content of the amino acids tyrosine and tryptophan, and this provides a sensitive and non-destructive form of assay. Proteins also absorb in the far ultraviolet because of the peptide bond.

Nucleic acids   Nucleic acids and their component bases show maximum absorption in the region of 260 nm. The extent of the absorption of nucleic acids is a measure of their integrity, since the partially degraded acids absorb more strongly than the native materials. The spectra of the component bases are also sufficiently different to be used in their identification.

Haemproteins   These conjugated proteins absorb in the visible region as well as in the UV region of the spectrum due to the haem group. When haemoglobin interacts with O_2, CO or drugs, characteristic shifts of these absorption maxima occur so the modified forms of the molecule can be detected and measured (Exp. 7.8). The visible spectra of the oxidized and reduced forms of cytochrome c are sufficiently different so that the relative amounts of these two forms can be determined in a mixture (Exp. 8.6).

Some practical points

The detailed operation of a particular instrument must, of course, be obtained by carefully reading the instruction manual, but a few general points concerning the use and care of colorimeters and spectrophotometers are given below.

1. *Cleaning cuvettes*   Cuvettes are cleaned by soaking in 50 per cent v/v nitric acid and then thoroughly rinsed in distilled water.
2. *Using the cuvettes*   First of all, fill the cuvettes with distilled water and check them against each other to correct for any small differences in optical properties. Always wipe the outside of the cuvettes with soft tissue paper before placing in the cell

holder and do not handle them by the optical faces. When all the measurements have been taken, wash them with distilled water and leave in the inverted position to drain.

3. *Absorption of radiation by cuvettes*   All cuvettes absorb radiation and the wavelengths at which significant absorption occurs depend on the material from which the cuvette is made (Fig. 7.8). Silica cuvettes are the most transparent to UV light but they are expensive so they are generally used in the far UV. Glass cuvettes are much cheaper than silica and so they are used whenever possible and invariably in the visible region of the spectrum. However, they do absorb UV and cannot be used below 360 nm. Plastic disposable cuvettes also absorb in the UV but much less than glass (Fig. 7.8) and they can be conveniently used at 340 nm for the assay of dehydrogenase enzymes that require NAD or NADP as coenzymes.

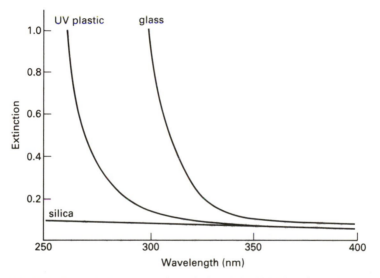

Fig. 7.8   The relative absorption of ultraviolet light by glass, plastic and silica cuvettes read against air as a blank

4. *Light source*   A tungsten lamp produces a broad range of radiant energy down to about 360 nm. To obtain the ultraviolet region of the spectrum a deuterium lamp is used as the light source. If the tungsten lamp is used in the range 360–400 nm, then a blue filter is placed in the light beam.

5. *Blanks*   The extinction of a solution is read against a reagent blank which contains everything except the compound to be measured. This blank is first placed in the instrument and the scale adjusted to zero extinction (100 per cent transmittance) before reading any test solutions. Alternatively, the extinction can be read against distilled water and the absorbance of the blank subtracted from that of the test solution.

6. *Replicates*   It is essential to prepare all blanks and standard solutions in duplicate so that an accurate standard curve can be constructed. In addition, the test solutions should also be prepared in duplicate wherever possible.

Experiments

Experiment 7.1   The absorbance curves of two dyes

PRINCIPLE

Coloured compounds have their own characteristic absorption spectra and careful selection of the wavelengths where maximum absorption is found enables a mixture of two coloured substances to be analysed.

MATERIALS	100
1. Colorimeter with a series of filters	50
2. Bromophenol blue (10 mg/litre)	1 litre
3. Methyl orange (10 mg/litre)	1 litre
4. An 'unknown' mixture of the two dyes	1 litre

METHOD

Determine the extinction of each dye in turn against the range of filters supplied with the colorimeter. Remember, the instrument must be reset on zero extinction with distilled water in the cuvette for each filter.

Carefully note the wavelength of maximum transmission (minimum absorbance) of each filter and plot a graph of the absorbance recorded against this wavelength.

What is the wavelength that gives maximum absorbance for each dye?

How does mixing the dyes affect the absorption spectrum?

Experiment 7.2   Demonstration of Beer's law

MATERIALS

1. As for Exp. 7.1.

METHOD

Prepare a range of concentrations of one of the dyes by setting up a series of tubes as below.

Bromophenol blue (10 mg/litre) (ml)	1	2	3	4	5
Distilled water (ml)	4	3	2	1	0

Place the filter which gave maximum extinction in the light path and zero the colorimeter with distilled water. Next record the absorbance of each solution and plot this against the concentration of dye in each tube (mg/litre or μg/ml).

1. Repeat the above experiment using the filter which gave maximum absorption with the methyl orange, one other filter, and, if possible, white light with no filter. How do the curves of extinction against concentration conform to Beer's law?

2. Repeat the whole of the experiment with the dye methyl orange.
3. Finally, use the information gained in these experiments to determine the concentration of each dye present in the mixture.

Experiment 7.3   The colorimetric estimation of inorganic phosphate

PRINCIPLE

Unlike the above dyes, most biochemical compounds are colourless and can only be analysed colorimetrically after reacting them with a specific chemical reagent to give a coloured product. This point is illustrated in the measurement of inorganic phosphate, which is probably one of the commonest determinations carried out in a biochemical laboratory, and the production of a standard curve now will prove useful in future experiments.

Inorganic phosphate reacts with ammonium molybdate in an acid solution to form phosphomolybdic acid. Addition of a reducing agent reduces the molybdenum in the phosphomolybdate to give a blue colour, but does not affect the uncombined molybdic acid. In this method the reducing agent used is *p*-methylaminophenol sulphate. The presence of copper in the buffer solution increases the rate at which the colour develops.

MATERIALS   <u>100</u>
1. Ammonium molybdate (50 g/litre)   1 litre
2. Copper acetate buffer pH 4.0. (Dissolve 2.5 g of copper sulphate   2 litres
   ($CuSO_4 \cdot 5H_2O$) and 46 g of sodium acetate ($CH_3COONa \cdot 3H_2O$) in
   1 litre of 2 mol/litre acetic acid. Check the pH and adjust to 4.0 if
   required)
3. Reducing agent. (Dissolve 20 g of *p*-methylaminophenol sulphate in   1 litre
   a 100 g/litre solution of sodium sulphite ($Na_2SO_3 \cdot 7H_2O$) and make
   up to 1 litre.) Store in a dark bottle until required
4. Trichloracetic acid (100 g/litre)   250 ml
5. Stock phosphate solution containing 100 mg phosphorus/100 ml.   100 ml
   (Dissolve 438 mg of potassium dihydrogen phosphate in water and
   make up to 100 ml.) Store in the refrigerator
6. Working phosphate solution containing 1 mg phosphorus/100 ml.   1 litre
   (Dilute the stock solution 1 in 100 with 50 g/litre TCA)
7. Colorimeter   50

METHOD

Pipette 0.1–1 ml of the standard solution of phosphate into a test tube, and, where necessary, add water to bring the final volume to 1 ml. Then add 3 ml of copper acetate buffer, 0.5 ml of ammonium molybdate, and 0.5 ml of reducing agent mixing thoroughly after each addition. Allow to stand for 10 min and read the extinction at 880 nm. Set up a blank by replacing the phosphate with 1 ml of 50 g/litre TCA.

Prepare a graph of the extinction against the concentration of phosphate.

When measuring the concentration of phosphate in a solution containing protein, the test solution is mixed with an equal volume of 20 g/litre TCA, the precipitate centrifuged, and an aliquot of the supernatant treated as above.

Experiment 7.4   The validity of Beer's law for the colorimetric estimation of creatinine

PRINCIPLE
Creatinine, in the presence of picric acid in alkaline solution, forms a red tautomer of creatinine picrate.

MATERIALS	100
1. Picric acid (saturated aqueous solution)	2 litres
2. Sodium hydroxide (1 mol/litre)	1 litre
3. Creatinine standard (2 g/litre)	1 litre
4. Colorimeter	50
5. Volumetric flasks (100 ml)	250

METHOD
Prepare a range of creatinine solutions by suitable dilution of the standard and place 1 ml of each solution in a standard 100 ml volumetric flask. Add 1 ml of 1 mol/litre sodium hydroxide and 2 ml of saturated picric acid solution to the flask, mix thoroughly, and stand for 10 min. Make up to the mark with water and measure the extinction at 530 nm.

Plot a graph of the extinction against the creatinine concentration.

Experiment 7.5   The absorption spectrum of p-nitrophenol

MATERIALS	10
1. p-Nitrophenol (10 mmol/litre)	50 ml
2. Hydrochloric acid (10 mmol/litre)	1 litre
3. Sodium hydroxide (10 mmol/litre)	1 litre
4. Spectrophotometer	5
5. Volumetric flasks (100 ml)	10

METHOD
Dilute the p-nitrophenol solution 0.2–50 ml with (a) 10 mmol/litre HCl and (b) 10 mmol/litre NaOH. Determine the absorption spectra of each solution from 250 to 500 nm. Comment on the differences between the two spectra and calculate the molar extinction coefficient at the wavelength for the maximum absorption.

Experiment 7.6   Determination of the pK_a value of p-nitrophenol

PRINCIPLE
p-Nitrophenol dissociates as shown in Fig. 7.9 and, as seen from the previous experiment, the undissociated form which is present in acid solution does not absorb

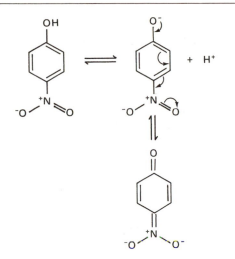

Fig. 7.9   The dissociation of *p*-nitrophenol

in the visible region while the quinonoid structure present in alkaline solution absorbs strongly. Now the pK value is the pH at which there is 50 per cent ionization or, in other words, the pH which gives half the absorbance value obtained in alkaline solution assuming 100 per cent ionization in alkali.

MATERIALS $\underline{10}$
1. As in the previous experiment   —
2. Buffer solutions, 50 mmol/litre   1 litre of each
    (a)   Citrate buffer (pH 3.0)   buffer
    (b)   Citrate buffer (pH 4.0)   solution
    (c)   Citrate buffer (pH 5.0)
    (d)   Citrate buffer (pH 6.0)
    (e)   Tris–HCl buffer (pH 7.0)
    (f)   Tris–HCl buffer (pH 7.5)
    (g)   Tris–HCl buffer (pH 8.0)
    (h)   Tris–HCl buffer (pH 9.0)
    (i)   Carbonate–bicarbonate (pH 10.0)
    (j)   Carbonate–bicarbonate (pH 11.0)

METHOD
Prepare 0.2 ml in 50 ml dilutions of the *p*-nitrophenol in the above buffer solutions and determine the extinction of each solution at a wavelength (405 nm) where the undissociated phenol has zero absorption. Determine the pK_a value. Titrate 25 ml of 10 mmol/litre *p*-nitrophenol with NaOH to a pH of 10.5. Correct the values obtained after titrating distilled water and prepare a titration curve. Determine the pK_a value and compare it with that obtained previously.

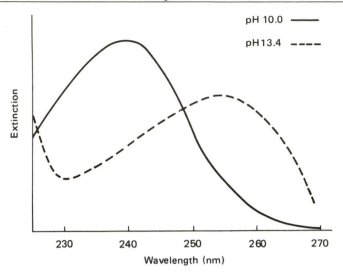

Fig. 7.10   The characteristic absorption spectra of barbiturates in alkaline solution

A third experiment involving *p*-nitrophenol could be carried out here, to show how these absorption properties can be used to follow an enzyme-catalysed reaction. Full details are given in Exp. 12.1.

Experiment 7.7   The estimation of barbiturates with the ultraviolet spectrophotometer

PRINCIPLE

The 5,5′-substituted barbiturates give a characteristic absorption spectrum in the ultraviolet with a maximum at 240 nm at pH 10. At pH 13.4 the maximum is shifted to 253 nm and a minimum is obtained at 235 nm. To detect the presence of barbiturates, the spectrum of a solution is plotted over the range 220–270 nm at pH 10.0 and pH 13.4 (Fig. 7.10). As well as the characteristic maxima and minima, the curves cross at 227 nm and 250 nm, the *isosbestic points*. There is also a maximum difference in absorption at 260 nm which is used to determine the amount of barbiturate present. The reason for this difference in absorption spectra is that these compounds are

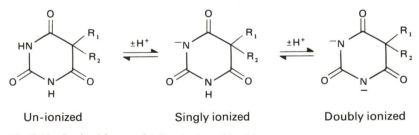

| Un-ionized | Singly ionized | Doubly ionized |

Fig. 7.11   Ionized forms of 5,5′-substituted barbiturates

dibasic acids with pK values of about 8 and 12, so that the singly ionized form is present at pH 10 and the doubly ionized at pH 13.4 (Fig. 7.11).

The detection and estimation of barbiturates in blood or urine is quite important in cases of suspected poisoning and the method can be used with blood, plasma, or urine. Barbiturates present in body fluids are extracted into chloroform, then into alkali.

An 'unknown' solution of phenobarbitone in plasma can be usefully examined. Concentrations in the plasma in excess of 100 mg/litre are usually fatal, and for convenience the 'unknown' should be somewhere in the region of 10–60 mg/litre.

MATERIALS		10
1. Chloroform (analytical grade)		1 litre
2. Na or K phosphate buffer (0.4 mol/litre, pH 7.4)		200 ml
3. Sodium hydroxide (50 mmol/litre)		200 ml
4. Boric acid–potassium chloride solution (37.2 g boric acid and 45 g potassium chloride in 1 litre)		100 ml
5. Phenobarbitone standard (20 mg/litre in 0.45 mol/litre NaOH)		100 ml
6. Whatman No. 31 filter papers		10
7. Separating funnels		20
8. Spectrophotometer		5
9. Sulphuric acid (6 mol/litre)		10 ml

METHOD

Extraction   Place 5–10 ml of the fluid in a separating funnel and extract three times with 30 ml of chloroform. Combine the extracts and filter through a Whatman No. 31 filter paper. Wash the filter with a little chloroform.

Return the extract to a clean separating funnel and wash twice with 5 ml of phosphate buffer, which is discarded. Add 10 ml of 0.45 mol/litre sodium hydroxide and shake vigorously for 1–2 min. Allow the phases to separate and centrifuge the aqueous layer.

Detection of barbiturate 1   Add 2 ml of the aqueous layer to 2 ml of 0.45 mol/litre sodium hydroxide and plot the spectra from 220 to 270 nm against a blank of 0.45 mol/litre sodium hydroxide (pH 13.4).

Add 2 ml of the aqueous layer to 2 ml boric acid solution. Measure this against a blank of 2 ml of 0.45 mol/litre sodium hydroxide and 2 ml of boric acid (pH 10.0).

Calculation 1   The *test* extinction is that of the pH 13.4 solution less that of the pH 10.0 solution at 260 nm.

The standard solution of phenobarbitone is treated the same as the aqueous layer from the chloroform extract and the *standard extinction* at 260 nm is calculated as for the test.

$$\text{Barbiturate conc. in fluid (mg/litre)} = \frac{\text{Test extinction}}{\text{Standard extinction}} \times 20$$

Detection of barbiturate 2   Record the spectrum at pH 10.0 as above then add four drops of 6 mol/litre H_2SO_4 to the sample and blank cuvettes and record the spectrum again, but this time at pH 2. Explain the difference observed and suggest what ionized form is probably present at pH 2. Treat the standard solution of phenobarbitone the same way and calculate the barbiturate concentration by using the difference in extinction at 240 nm of the two solutions. Which of the two methods is the more sensitive?

Experiment 7.8   Experiments with haemoglobin

PRINCIPLE

When haemoglobin combines with oxygen, there is a shift in the absorption spectrum and the colour of the blood is changed from dark to bright red; the reverse change occurs on deoxygenation and this is the reason for the difference in colour of venous and arterial blood. In both haemoglobin and oxyhaemoglobin, the iron is present in the ferrous form and is not oxidized on oxygenation. If the ferrous iron is oxidized to ferric with an oxidizing agent such as ferricyanide, then methaemoglobin is formed and the molecule can no longer combine with oxygen or carbon monoxide. Normally, human blood contains only about 1 per cent methaemoglobin, but this may be increased following the ingestion of certain drugs.

Haemoglobin combines with carbon monoxide some 200 times more readily than with oxygen to form carboxyhaemoglobin; the amount of haemoglobin available for oxygen transport is thereby reduced and, if sufficient CO is present, death ensues from oxygen starvation of the tissues. The absorption spectrum of carboxyhaemoglobin is only slightly different from that of haemoglobin, but the difference is sufficient to be the basis for detecting the compound in blood and this may have medico-legal implications in cases of death from coal gas poisoning.

If blood is shaken in the air then oxyhaemoglobin is formed, so Stokes' reagent is added to the blood to remove the oxygen completely and form haemoglobin.

Haemoglobin derivative	Absorption maxima (nm)		
Haemoglobin	555	430	—
Oxyhaemoglobin	577	541	413
Carboxyhaemoglobin	570	535	418
Methaemoglobin	630	500	406

MATERIALS   10
1. Ultraviolet spectrophotometer   5
2. Stokes' reagent (20 g/litre ferrous sulphate and 30 g/litre tartaric acid: 50 ml
   just before use add ammonia until a faint precipitate which forms at

first is just dissolved. This is a solution of ammonium ferrotartrate, a reducing agent)

3. Potassium ferricyanide (100 g/litre)	50 ml
4. Haemoglobin preparation from haemolysed red cells of human blood	50 ml
5. Sodium chloride (0.15 mol/litre)	200 ml
6. Direct vision spectroscope	2
7. Source of carbon monoxide	1

METHOD

Prepare a solution of haemoglobin in saline at a concentration of about 1 mg/ml and plot the absorption spectrum over the range 400–700 nm.

Plot the absorption spectrum after treating the haemoglobin solution as below:

1. Add two drops of freshly prepared Stokes' solution to 4 ml of the haemoglobin solution.
2. Pass carbon monoxide through the haemoglobin solution (not in the open lab) and seal the top of the cuvette.
3. Repeat (2) after adding two drops of Stokes' reagent.
4. Add two drops of 100 g/litre potassium ferricyanide solution to 4 ml of haemoglobin.
5. Repeat (4), but add Stokes' reagent after the ferricyanide.

In addition, examine the visible spectra of the above solutions using a direct vision spectroscope.

Experiment 7.9   The ultraviolet absorption of proteins and amino acids

PRINCIPLE

Absorption at 210 nm   Below 230 nm, the extinction of a protein solution rises steeply reaching a maximum at 190 nm; this is mainly due to the peptide bond. In practice, it is more convenient to measure the extinction at 210 nm where the specific extinction coefficient $E_{1\,cm}^{10\,g/l}$ is about 200 for most proteins. All proteins have a similar specific absorption here since the peptide bond content is similar.

A number of compounds such as carboxylic acids, buffer ions, alcohols, bicarbonate and aromatic compounds also absorb in this region, so the results need to be interpreted with care.

Absorption at 280 nm   Tyrosine and tryptophan absorb at 275 nm and 280 nm and so proteins containing these amino acids will also absorb in this region. The specific extinction coefficient $E_{1\,cm}^{10\,g/l}$ varies according to how much of these amino acids is present in the particular protein. The values found in practice range from 6 to 60, although many proteins have a value close to 10; that is 1 mg/ml of protein gives an extinction at 280 nm of about 1 when viewed through 1 cm light path.

The disadvantage of this method is that many other compounds absorb in this region, particularly nucleic acids which have a peak at 260 nm. Pure proteins have a ratio of absorption (at 280 nm/260 nm) of about 1.8, while nucleic acids have a ratio of 0.5.

MATERIALS	<u>10</u>
1. Proteins (5 g/litre albumin and casein)	50 ml
2. Amino acids (0.1 mmol/litre tyrosine, tryptophan, and phenylalanine in water adjusted to pH 7, in 10 mmol/litre HCl and 10 mmol/litre NaOH)	50 ml
3. Ultraviolet spectrophotometer	5

METHOD

Plot the absorption spectra of the above compounds over the range 190–400 nm. At the lower end of the wavelength, the proteins will need to be diluted about 1 in 200. What amino acids absorb the strongest at 280 nm?

Fluorescence spectroscopy

Fluorescence

Some compounds not only absorb radiation but also emit some of the energy in the form of fluorescent light. Energy is absorbed in the UV region of the spectrum and molecules are elevated from the ground state to a high energy level. The excited molecules then return to the ground state with the consequent emission of visible light. The wavelength of the emitted light is always higher than that of the absorbed radiation.

The requirements for a compound to fluoresce are an absorbing structure and a high resonance energy. Aromatic compounds in general are often capable of fluorescence, particularly if the substituent in the ring is electron donating.

Quenching

However, fluorescence is not so common as absorption due to *quenching*. Molecules containing Br, I, NO_2 and azo groups show little fluorescence because of this. Quenching decreases the quantum yield so that the absorbed energy is used for competitive electronic transitions with excited molecules or for breaking weak bonds instead of being emitted as fluorescent light. Quenching can also occur by interaction with the solvent and other molecules in solution. In some cases, the quenching reactions are fairly specific and can be used to identify a particular fluorescent compound.

Applications

Fluorescent compounds are used extensively in biochemical investigations as they can be detected at very low concentrations and with a high degree of selectivity. The absorption and fluorescent spectra of a compound are quite characteristic so that, when the maxima are selected by filters or monochromators on the incident and emitted beams, the fluorescent compound can be detected and measured even when other fluorescent compounds are present.

Some of the applications include the use of fluorescent compounds as membrane probes, substrates for sensitive enzyme assays and immunofluorescence.

Intensity of fluorescence and concentration

The fluorescence (F) depends on the intensity of light absorbed, so that if the intensity of the incident and emergent beams is I_0 and I, respectively, then:

$$F = K(I_0 - I)$$

Now Beer's law states that:

$$I = I_0 e^{-kcl}$$

so that,

$$I_0 - I = I_0(1 - e^{-kcl})$$

therefore,

$$F = KI_0(1 - e^{-kcl})$$

Expanding the exponential expression and assuming c to be small so that the higher terms can be ignored,

$$F = KI_0(kcl)$$

The constant K is known as the quantum yield and is the ratio of the number of quanta emitted to the number absorbed. For a particular compound and instrument

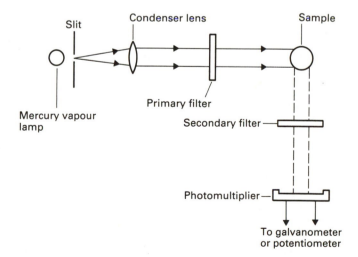

Fig. 7.12 A diagram of the general arrangement of a fluorimeter

when I_0, k and l are constant, the fluorescence is directly proportional to concentration:

$$F = K'c$$

This equation holds in practice providing the solution absorbs less than 5 per cent of the exciting radiation so that, the greater the intensity of the incident light, the higher is the concentration that gives a linear response. High light intensities also produce problems such as photodecomposition and light scattering. The useful concentration range for the determination of fluorescent compounds is 0.001–10 µg/ml depending on the material under investigation.

Fluorimeters

A diagram of a typical arrangement of a fluorimeter is shown in Fig. 7.12.

Light source   A high intensity source of ultraviolet light is provided by a mercury vapour lamp. Ultraviolet light is a hazard to be reckoned with (Chapter 1) so adequate shielding of the lamp must be provided.

Filters   After passing through a condenser lens, the light meets the *primary filter*, which isolates a particular region of the ultraviolet spectrum, then on through the sample in a circular cuvette to the secondary filter set at right angles to the incident beam. The *secondary filter* cuts off all the scattered light and passes all the emitted fluorescence. The filters are carefully selected so that their transmission matches the excitation and fluorescent spectra as closely as possible. Ideally, monochromatic light should be used but this increases the cost of the instrument. A fluorimeter with a monochromator on the secondary side and filters on the primary side is adequate for many purposes.

Photomultiplier   The light then passes on to the photomultiplier which produces an electrical signal proportional to the intensity of the fluorescent beam. This may then be displayed or recorded on an appropriate measuring device such as a microammeter or flat bed recorder.

Experiment 7.10   The sensitivity of fluorescence assays

PRINCIPLE

4-Methylumbelliferyl compounds form the basis of a number of fluorimetric assays of hydrolases (Fig. 7.13). The product 4-methylumbelliferone is highly fluorescent and can be distinguished from the lower fluorescence of the substrate by careful selection of the excitation and emission wavelengths.

The sensitivity of such assays is demonstrated by comparing the lowest concentration of 4-methylumbelliferone that can be detected by absorption spectroscopy with the concentration of the compound that can be detected by fluorimetry.

Quinine sulphate is included so that maximum reading is obtained with 4 µg/ml.

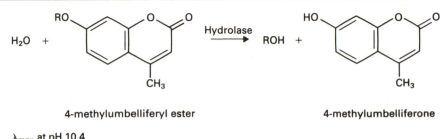

4-methylumbelliferyl ester   4-methylumbelliferone

λ_{max} at pH 10.4

Absorption	320 nm	360 nm
Fluorescence	380 nm	450 nm

Fig. 7.13   The use of fluorescent substrates for the assay of hydrolases

MATERIALS   10
1. Fluorimeter   5
2. UV spectrophotometer   5
3. Quinine sulphate (4 µg/ml in 0.1 mol/litre H_2SO_4)   50 ml
4. 4-Methylumbelliferone primary standard (1 mmol/ml in methanol)   10 ml
5. 4-Methylumbelliferone (0.1 mmol/litre, prepared by diluting the   50 ml
   primary standard with glycine buffer)
6. Glycine buffer (0.05 mol/litre, pH 10.4)   100 ml
7. H_2SO_4 (0.1 mol/litre)   100 ml

METHOD

Absorption   Plot the absorption spectra against the solvent blank for the 0.1 mmol/litre 4-methylumbelliferone and the quinine sulphate. Determine the wavelength that gives the maximum absorption and measure the extinction of a range of concentrations of each compound. Prepare a plot of extinction against concentration for the quinine sulphate and the 4-methylumbelliferone.

Fluorescence   Set the fluorimeter to give a primary wavelength of 360 nm and a secondary wavelength of 420 nm and adjust the galvanometer to give a maximum deflection with the solution of quinine sulphate. Prepare a series of dilutions down to 0.01 µg/ml of the quinine sulphate and 5 ng/ml of the 4-methylumbelliferone. Read the fluorescence of each solution and plot a graph of fluorescence against concentration.

Compare the relative sensitivities of the absorption and fluorescence assays of the two solutions and estimate the minimum concentration that can be assayed by the two methods.

Experiment 7.11   Fluorescence quenching

MATERIALS   10
1. Solutions for Exp. 7.10   —

2. 2:4 Dinitrophenol (0.1 mmol/litre in 0.05 mol/litre glycine buffer, pH   50 ml
   10.4)
3. HCl (0.1 mol/litre)   50 ml

METHOD
1. Repeat the standard curve of 4-methylumbelliferone in glycine buffer but this time include 0.1 mmol/litre 2:4 DNP in the solutions: explain the result.
2. Prepare a standard curve of the quinine sulphate in 0.1 mol/litre 2:4 DNP.

Experiment 7.12   The measurement of α-naphthyl phosphatase activity

PRINCIPLE
At pH 10, alkaline phosphatase hydrolyses α-naphthyl phosphate to α-naphthol and phosphate. The solution is then adjusted to pH 12 with sodium hydroxide when the α-naphthol is present largely as the phenolate ion which is highly fluorescent (Fig. 7.14).

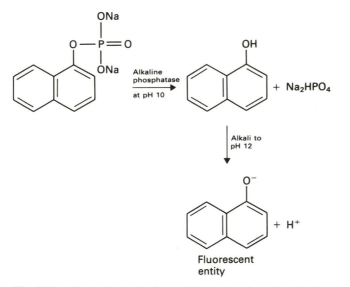

Fig. 7.14   The hydrolysis of α-naphthyl phosphate by alkaline phosphatase

The substrate is slightly fluorescent but careful selection of the excitation and emission wavelengths avoids the weak fluorescence of the α-naphthyl phosphate.

MATERIALS   10
1. α-Naphthol (1 mmol/litre prepared by dissolving the pure compound   200 ml
   in a little ethanol and making up to the mark with water)
2. α-Naphthyl phosphate (10 mmol/litre, stored in a brown bottle)   50 ml
3. Sodium carbonate–bicarbonate buffer (0.1 mol/litre, pH 10)   100 ml

4. Sodium hydroxide (0.5 mol/litre)   100 ml
5. Fluorimeter: 5
      Primary wavelength 335 nm
      Secondary wavelength 455 nm
6. Alkaline phosphatase   —

METHOD

Standard curve   Make up a range of concentrations of α-naphthol from 0.1 μmol/litre to 1 μmol/litre and prepare the following mixtures:

	1 (ml)	*2* (ml)	*3* (ml)
α-Naphthol (0.1 μmol/litre – 1 μmol/litre)	1	1	1
Sodium carbonate–bicarbonate buffer	1	1	—
Water	1	—	3
Sodium hydroxide (0.5 mol/litre)	1	1	—
α-Naphthyl phosphate (10 mmol/litre)	—	1	—

Set the potentiometer reading to 100 for the highest concentration of α-naphthol in 1 and prepare standard curves of the potentiometer readings for each concentration of α-naphthol in 1, 2 and 3.
What effect does the α-naphthol have and why? (Tube 2)
What happens when the α-naphthol is prepared in water? (Tube 3)

Enzyme activity   Prepare a mixture of 1 ml of α-naphthyl phosphate and 1 ml of buffer solution, equilibrate at 37°C and add 1 ml of an aqueous solution of alkaline phosphatase of suitable dilution. After a measured time interval, add 1 ml of 0.5 mol/litre sodium hydroxide to stop the reaction and give the pH for maximum fluorescence of the α-naphthol. Read the fluorescence in the fluorimeter and plot a progress curve of the reaction by stopping the reaction after different intervals of time. Express the activity from the initial reaction rate as μmoles of α-naphthol produced per ml of enzyme.

Calculations

1. A solution of 10^{-5} mol/litre ATP shows a transmittance of 70.2 per cent at 260 nm in a 1 cm cuvette, calculate:
   (a) the absorbance;
   (b) the transmittance in a 3 cm cuvette;
   (c) the absorbance of 50 μmol/litre ATP in a 1 cm cuvette.
2. The specific extinction coefficient of a glycogen–iodine complex at 450 nm is 0.20.

Calculate the concentration of glycogen in a solution of iodine which has an extinction of 0.36 in a 3 cm cuvette.

3. A solution of UTP of 29.3 mg/litre has an extinction of 0.25 at 260 nm. If the light path is 1 cm and the molecular weight of the UTP is 586, calculate:
   (a) the molar extinction coefficient;
   (b) the transmittance of a 10 µmol/litre solution.

4. A solution of the amino acids tyrosine and tryptophan has an extinction of 0.65 at 280 nm and 0.5 at 295 nm in a 1 cm cuvette. Given the extinction coefficients of the pure amino acids, calculate the concentration of tyrosine and tryptophan present in the mixture.

| Wavelength (nm) | Molar extinction coefficient (litres mol^{-1} cm^{-1}) | |
	Tyrosine	Tryptophan
280	1500	5000
295	2500	2500

5. A tissue extract (0.3 ml) was diluted with 0.9 ml of water. An aliquot of the diluted solution (0.5 ml) was added to 2.5 ml of biuret reagent and gave an extinction of 0.324 at 540 nm in a 1 cm cuvette. A standard solution of albumin (4 mg/ml) gave an extinction of 0.24 when 0.5 ml was added to 2.5 ml of the biuret reagent. Calculate the concentration of protein in the original tissue extract.

6. A solution containing NAD^+ and NADH has an extinction of 0.316 at 340 nm and 1.11 at 260 nm. Calculate the concentrations of oxidized and reduced forms of the coenzyme in the solution given that both NAD^+ and NADH absorb at 260 nm but only NADH absorbs at 340 nm.

| Wavelength (nm) | Molar extinction coefficient (litres mol^{-1} cm^{-1}) | |
	NAD^+	NADH
260	18 000	15 000
340	0	6 320

7. A fluorescent compound gives a reading of 10 in a fluorimeter at a concentration of 20 µmol/litre. If the cuvette has a light path of 1 cm and the compound an extinction coefficient of 5270, what fluorescent reading would be expected with a concentration of 40 µmol/litre? How much does this value differ from the expected reading of 20 which assumes that the fluorescence varies linearly with concentration?

Further reading

Campbell, I. D. and Dwek, R. A., *Biological Spectroscopy*. Benjamin/Cummings, Menlo Park, 1984.

Udenfriend, S., *Fluorescence Assay in Biology and Medicine*. Academic Press, New York, 1969.

Van Holde, K. E., *Physical Biochemistry*, 2nd edn. Prentice-Hall, Englewood Cliffs, 1985.

Burrin, D. H., 'Spectroscopic Techniques' in *A Biologist's Guide to Principles and Techniques of Practical Biochemistry*, K. Wilson and K. H. Goulding (eds). 3rd edn. Arnold, London, 1986.

Wrigglesworth, J. M., *Biochemical Research Techniques: A Practical Introduction*. Ellis Horwood, Chichester, 1983.

SECTION THREE   Molecules of the biosphere

8. Amino acids and proteins

Chemical and physical properties

Chemistry of the amino acids

Acid–base properties   As the name suggests, amino acids are organic compounds that contain amino and carboxyl groups, and therefore possess both acidic and basic properties. There are a large number of chemically possible amino acids, but only a few of these occur naturally. In the case of the 22 or so amino acids found in proteins, nearly all of them are α-amino acids, where the amino group is present on the α-carbon atom.

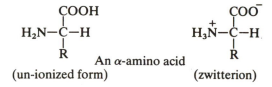

An α-amino acid

(un-ionized form)   (zwitterion)

Amino acids are unlike low molecular weight organic compounds in their properties and resemble inorganic salts. In general, they are readily soluble in aqueous media, but only slightly soluble or insoluble in organic solvents. Their melting points are also very high for low molecular weight organic compounds. This is because amino acids exist mostly as *zwitterions* and not as un-ionized molecules. The high melting points are due to the high energy required to break the ionic bonds of the crystal lattice.

The strong positive charge on the $-NH_3^+$ group induces a tendency for the $-COOH$ group to lose a proton, so that amino acids are strong acids. The pK_a for glycine for example is much lower than the corresponding aliphatic acid as seen below.

	$\overset{H}{\underset{}{\overset{|}{{}^+H_3N-CH-COO^-}}}$	$\overset{H}{\underset{}{\overset{|}{H-CH-COOH}}}$	$\overset{H}{\underset{}{\overset{|}{H_2N-CH-H}}}$
	Glycine	*Acetic acid*	*Methylamine*
Physical state at 25°C	Solid	Liquid	Gas
Melting point	232–236°C	17°C	−94°C
pK —COOH	2.4	4.8	—
pK —NH$_2$	9.7	—	10.7

Some amino acids contain ionizable groups in the side chain R, and these affect their physical characteristics whether the amino acids are free in solution or combined with others in a protein. In fact, the charge properties of proteins are determined to a large extent by the ionizable groups of the amino acid side chains. The ionizable groups present in the side chains of amino acids, together with their pK values, are given in Table 8.1.

Isoelectric point   Amino acids migrate in an electric field and this property is the basis of one method for their separation. The direction and extent of migration depend to a large extent on the predominant ionic form present and this is determined by the pH of the electrophoresis buffer (Exp. 5.1).

The pH at which there is zero net charge and no migration in an electric field is known as the *isoelectric point*. For amino acids containing only one —COOH and one —NH$_2$ as the ionizable groups, the isoelectric point (pI) is half-way between the pK values of these groups. So that in the case of alanine:

$$pI = \tfrac{1}{2}(2.4 + 9.7) = 6.1$$

Ionized forms of alanine	HOOC—CH—NH$_3^+$ (CH$_3$)	$^-$OOC—CH—NH$_3^+$ (CH$_3$)	$^-$OOC—CH—NH$_2$ (CH$_3$)
pH where they exist	Acid	Isoelectric	Alkaline
Charge carried	+	0	−
Migration in electric field	To cathode	None	To anode

When other charged groups are present, the calculation of the pI is not so simple, but, as a rough guide, the isoelectric point lies midway between the pK values of similar groups.

Stereochemistry   The α-carbon atom is *asymmetric* and is a *chiral centre* for all amino acids except glycine so that, apart from glycine, all amino acids show optical activity. If serine is taken as the parent compound, then this can be compared with L(—) glyceric acid, the parent compound for the L series of sugars.

COOH	COOH	COOH
HO—C—H	H$_2$N—C—H	H—C—NH$_2$
CH$_2$OH	CH$_2$OH	CH$_2$OH
L(−) Glyceric acid	L(+) Serine	D(−) Serine

When —CH$_2$OH is replaced by other groups, two families of amino acids emerge, the D and L series. All of the amino acids present in proteins are of the L configuration although the D form is found in antibiotics and bacterial cell walls. For a fuller

Table 8.1   The pK values of ionizable groups found in the side chains of some amino acids

Ionizing group	Amino acid	pK
β-Carboxyl	Aspartic acid	3.9
γ-Carboxyl	Glutamic acid	4.3
$-COOH \rightleftharpoons -COO^- + H^+$		
Imidazole	Histidine	6.0

$$^+HN\diagup\!\!\!\!\diagdown\!\overline{N}H \rightleftharpoons N\diagup\!\!\!\!\diagdown\!\overline{N}H + H^+$$

Sulphydryl	Cysteine	8.3
$-CH_2SH \rightleftharpoons -CH_2S^- + H^+$		
Phenolic	Tyrosine	10.1

$$OH \qquad O^-$$
$$\bigcirc \qquad \bigcirc + H^+$$

ε-Amino	Lysine	10.5
$-NH_3^+ \rightleftharpoons NH_2 + H^+$		
Guanidino	Arginine	12.5

$$\begin{array}{ccc} NH_2 & & NH_2 \\ | & & | \\ C=NH_2^+ & \rightleftharpoons & C=NH + H^+ \\ | & & | \\ NH & & NH \end{array}$$

explanation of these terms, and more information on optical isomerism, please read the relevant section in the introduction to Chapter 9.

The L and D refer to the absolute configuration about the chiral centre and not to the optical activity.

Some amino acids such as isoleucine, threonine, and hydroxylysine contain a second chiral centre so that more than two forms are possible.

The amino acid composition of proteins

Formulae of amino acids   Just as monosaccharides are the basic unit of polysaccharides, so amino acids can be thought of as the 'bricks' from which the protein 'house' is built. Polysaccharides are built up of usually only a few monosaccharide units, but proteins may contain as many as 22 different amino acids. The names and abbreviations of the amino acids commonly found in proteins are given in Table 8.2. The formulae of all the α-amino acids are the same except for the nature of the side chain R.

Peptide bond   The amino acids are joined together in the protein molecule by peptide bonds (—CO—NH—) formed by the condensation of the α-COOH of one amino acid with the α-NH$_2$ group of another one. Low molecular weight polymers of the amino

Table 8.2 The common amino acids found in proteins

Group	Name	Abbreviation	R
Aliphatic	Glycine	Gly	$H-$
	Alanine	Ala	CH_3-
	Valine	Val	$\begin{array}{c} CH_3 \\ | \\ CH- \\ | \\ CH_3 \end{array}$
	Leucine	Leu	$\begin{array}{c} CH_3 \\ | \\ CH-CH_2- \\ | \\ CH_3 \end{array}$
	Isoleucine	Ileu	$\begin{array}{c} CH_3 \\ | \\ CH- \\ | \\ C_2H_5 \end{array}$
Hydroxylic	Serine	Ser	CH_2OH
	Threonine	Thr	$\begin{array}{c} CH_3 \\ | \\ CHOH \end{array}$
Sulphur	Cysteine	Cys	CH_2SH
	Cystine	Cys-Cys	$CH_2-S-S-CH_2$
	Methionine	Met	$\begin{array}{c} CH_2 \cdot S \cdot CH_3 \\ | \\ CH_2 \end{array}$
Acidic	Aspartic acid	Asp	$\begin{array}{c} COOH \\ | \\ CH_2 \end{array}$
	Glutamic acid	Glu	$\begin{array}{c} COOH \\ | \\ CH_2 \\ | \\ CH_2 \end{array}$
Basic	Lysine	Lys	$\begin{array}{c} NH_2 \\ | \\ (CH_2)_4 \end{array}$
	Arginine	Arg	$\begin{array}{c} HN \quad NH_2 \\ \diagdown C \diagup \\ | \\ NH \\ | \\ (CH_2)_3 \end{array}$
Aromatic and heterocyclic	Phenylalanine	Phe	CH_2- (phenyl ring)
	Tyrosine	Tyr	CH_2- (phenol ring with OH)

Table 8.2   (continued)

Group	Name	Abbreviation	R
	Tryptophan	Try	
	Histidine	His	
Imino acids	Proline	Pro	
	Hydroxyproline	Hyp	

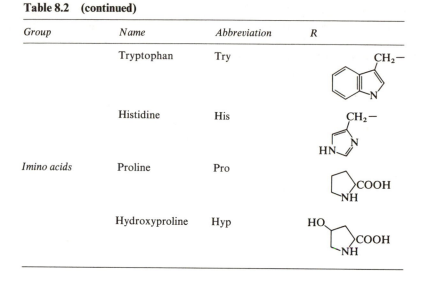

acids are known as *polypeptides*, while the term proteins is usually reserved for the larger polymers of molecular weight several thousand or more.

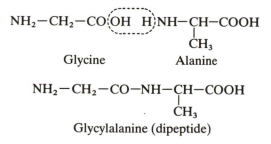

$$NH_2-CH_2-CO[OH \quad H]NH-CH-COOH$$
$$\qquad\qquad\qquad\qquad\qquad\qquad CH_3$$

Glycine   Alanine

$$NH_2-CH_2-CO-NH-CH-COOH$$
$$\qquad\qquad\qquad\qquad\qquad CH_3$$

Glycylalanine (dipeptide)

The —C—N— atoms of the peptide bond all lie in the same plane to form the backbone, and the side chains of the individual amino acids are arranged *trans* to each other across the backbone.

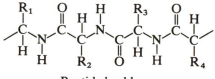

Peptide backbone

Protein structure

Primary structure   Several levels of structural organization can be recognized in proteins and the first of these is the primary structure, which is the sequence of amino acids. The constituent amino acids of a pure protein can be readily determined by separating them by chromatography or electrophoresis after chemical or enzymic hydrolysis of the peptide bonds.

The quantitative analysis of the protein by automatic ion exchange chromatography is fairly straightforward but trying to place the hundreds or thousands of amino acid residues in the right order is more difficult. Fortunately, chemical methods are available for the identification of the free —COOH and —NH$_2$ groups of proteins and peptides, and this provides the key to the problem.

The protein is partially hydrolysed by acid, which gives random hydrolysis, and by the use of enzymes, which catalyse the hydrolysis of peptide bonds between specific amino acids. A large number of peptides are obtained which are separated by chromatography and electrophoresis and the C- and N-terminal groups identified. The structures of the di- and tripeptides are determined. These are then used to elucidate the amino acid sequence of the large peptide fragments until the complete sequence is known, rather like constructing a jigsaw puzzle.

Secondary structure   Pauling and Corey, on the basis of X-ray studies, suggested that the peptide chain can exist in the form of a coil or helix. A number of helical forms were considered by these workers, but only the *α-helix* met all the requirements for maximum stability. This helical form has 3.6 amino acid residues per complete turn and a rise along the central axis of 0.15 nm per residue. The shape of the structure is maintained by intramolecular hydrogen bonds between the carbonyl oxygen and the amide nitrogen three residues apart in the peptide backbone:

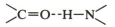

The hydrogen bond is fairly weak, but the large number involved in the formation of the α-helix maintains the structure in this stable form. The amino acid side chains can be accommodated in the α-helix since these 'stick out' into space away from the coil. The imino acids proline and hydroxyproline, however, do not fit into the normal α-helix and, where these are present, a kink or change in direction of the chain occurs. This is because of the rigid nature of the structure of these imino acids.

Another form of secondary structure is the *β* pleated sheet, where hydrogen bonding is between two peptide chains which may be *parallel*, in that their N atoms point in the same direction, or *antiparallel*, where alternate chains are orientated the same way. The *β* form is found in fibrous proteins such as hair keratin, while the α-helix may be present in both fibrous and globular proteins like albumin and myoglobin.

Tertiary structure   The tertiary structure of a protein is the arrangement in space of the molecular threads or, in other words, the overall shape of the protein molecule. Many protein molecules behave as if they are very compact and are therefore known as *globular proteins*. Other proteins are more rigid and form long thin threads, these are the *fibrous proteins*. The tertiary structure is maintained by a number of bonds of the type shown below.

The covalent disulphide bond formed between cysteine residues is the strongest and confers a certain amount of rigidity on the protein.

The ionic bond occurs when an ionized acidic and basic group are brought into close proximity and this bond of moderate strength is important in the binding of basic proteins with acidic macromolecules, as in the formation of nucleoproteins.

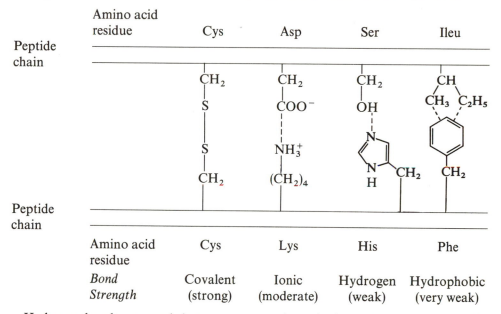

Amino acid residue	Cys	Lys	His	Phe
Bond Strength	Covalent (strong)	Ionic (moderate)	Hydrogen (weak)	Hydrophobic (very weak)

Hydrogen bonds are weak but numerous and can be formed between the amide nitrogen and the carbonyl oxygen of the peptide backbone as well as groups present in the side chain. The side chains of aspartic acid, glutamic acid, tyrosine, histidine, serine and threonine are all capable of hydrogen bond formation.

Hydrophobic bonds arise from a tendency for the non-polar side chains of amino acid residues to associate with each other. The hydrophilic groups are associated with water and are found on the surface of the protein molecule. This type of association dictates the folding of the polypeptide chain and therefore the overall shape of the molecule.

Quaternary structure   A number of proteins are made up of polypeptide units which are not covalently linked and the association of these subunits to form the molecule confers quaternary structure on the protein. Probably the best known example of this is haemoglobin which consists of 2α and 2β subunits.

Function in the living organism

The name protein is derived from the Greek *proteios* meaning 'of primary importance', and the name is fully justified. The main function of proteins is to act as essential components of structural material, in contrast to that of carbohydrates and fats which is to provide energy. This does not mean that proteins are static compounds; on the contrary, they are in a state of continuous flux with regard to their synthesis and degradation.

Amino acids

The amino acids found in proteins arise mainly from the digestion of dietary proteins. Some amino acids can be synthesized by the animal and these are known as *non-essential* amino acids, while others, the *essential* amino acids, must be supplied in the diet.

The α-amino acids, as well as being concerned in protein synthesis, are also involved

Table 8.3 Some important amino acids and derivatives

Name and formula	Occurrence and role
γ-Aminobutyric acid $COOH$ $(CH_2)_3$ NH_2	This is formed in the brain by decarboxylation of glutamic acid, where it may act as a chemical mediator in the transmission of the nerve impulse between some neurones.
Asparagine $COOH$ $CHNH_2$ CH_2CONH_2	Asparagine is present in a number of plant tissues, where it acts as a reservoir of nitrogen for protein synthesis during germination.
Diaminopimelic acid $COOH$ $CHNH_2$ $(CH_2)_3$ $CHNH_2$ $COOH$	This amino acid is an important constituent of the mucopeptide of bacterial cell walls.
3,4-Dihydroxyphenylalanine $CH_2CH(NH_2)COOH$	This amino acid, known as DOPA for short, is a precursor of the important pigment melanin. This pigment is responsible for the colour of hair, skin and eyes.
Histamine $CH_2CH_2NH_2$	Histamine arises from histidine by decarboxylation. It is a vasodilator and is involved in shock and allergic responses.

in the synthesis of a number of compounds of biological importance, some of which are given in Table 8.3.

Peptides

The term peptide is generally used for polymers containing up to 50 amino acid residues or with a molecular weight of less than 5000. Peptide chains are found attached to carbohydrate material to form the peptidoglycans of bacterial cell walls and the glycoprotein of blood group substances. In the former case, many of the amino acids are of the D and not the usual L configuration.

A number of peptides are also found in the free state and are probably intermediates in the turnover of proteins, although some peptides are synthesized to perform specific biological functions (Table 8.4).

Table 8.4   Some examples of biologically important peptides

Peptide	Structure
Glutathione	Glu–Cys–Gly

This is a tripeptide with one of the peptide bonds formed through the γ- rather than the α-carboxyl group of glutamic acid. Its main function is to protect the thiol groups of molecules and membranes by keeping them in the reduced state.

Gramicidin S	D-Phe–L-Leu–L-Orn–L-Val–L-Pro
	| |
	L-Pro–L-Val–L-Orn–L-Leu–D-Phe

This cyclic decapeptide is an antibiotic from the bacterium *Bacillus brevis* and contains the unusual amino acids D-phenylalanine and L-ornithine. A number of other antibiotics are also peptides.

Oxytocin	┌────S────────S────┐
	Cys–Tyr–Ileu–Gln–Asn–Cys–Pro–Leu–Gly

This peptide is a hormone synthesized in the posterior lobe of the pituitary and causes uterine contraction and ejection of milk in the female animal.

Proteins

An adequate intake of protein is essential for higher animals since only the simple forms of life are able to synthesize their protein from other nitrogen sources. Proteins are present in all tissues of the body and make up a large part of the structure of the cell. In addition, a number of proteins have specialized physiological roles.

The numbers and types of proteins present in living matter are vast and only a few can even be briefly considered.

Membranes   The membranes of all cells and organelles contain protein in association with lipid. Lipoprotein membranes act as selective permeability barriers and are involved in the transport of materials into and out of the cell and its compartments.

A number of enzymes, including those involved in the biosynthesis of macromolecules and the detoxication of foreign compounds, are also part of the cell membrane of some tissues.

The plasma proteins   Blood plasma contains large amounts of a number of proteins, each of which has a specialized biological function. Plasma *albumin*, for example, acts as a store of protein, is important in the maintenance of plasma pH and osmotic pressure, and transports a variety of compounds in the blood. The *α- and β-globulins* are associated with the transport of lipids and the *γ-globulins* with antibodies.

Fibrinogen is a soluble protein which is converted into insoluble *fibrin* during blood clotting.

Hormones   Several proteins have hormonal properties. *Insulin*, secreted by the *β* cells of the pancreas, controls carbohydrate metabolism by lowering the blood sugar, while glucagon, from the *α* cells, increases the blood sugar. *Gastrin* stimulates acid secretion in the stomach and the *parathyroid hormone* is concerned in the regulation of calcium and phosphate metabolism.

Enzymes   All enzymes are proteins, and the importance and wide occurrence of these biological catalysts are well known. They owe their high specificity and catalytic activity to the correct arrangement of amino acids in space for the substrate to bind to.

Assay methods

Experiment 8.1   The quantitative estimation of amino acids using the ninhydrin reaction

PRINCIPLE
Ninhydrin (triketohydrindene hydrate) reacts with α-amino acids between pH 4 and 8 to give a purple-coloured compound. Not all amino acids give exactly the same intensity of colour and this must be allowed for in any calculation. The amino acids proline and hydroxyproline give a yellow colour, so these are read at 440 nm.

MATERIALS	100
1. Amino acids (0.1 mmol/litre aspartic acid, arginine, leucine, and proline)	1 litre
2. Acetate buffer (4 mol/litre, pH 5.5)	500 ml
3. Methyl cellosolve (ethylene glycol monomethyl ether)	1 litre
4. Ethanol (50 per cent v/v)	2 litres
5. Ninhydrin reagent. (Dissolve 0.8 g of ninhydrin and 0.12 g of hydrindantin in 30 ml of methyl cellosolve and add 10 ml of acetate buffer; prepare fresh and store in a brown bottle. *Care*: carcinogenic!)	1.2 litres

METHOD

Pipette 2 ml of the amino acid solution into a test tube, add 2 ml of the buffered ninhydrin reagent, and heat in a boiling water bath for 15 min. Cool to room temperature, add 3 ml of 50 per cent ethanol, and read the extinction at 570 nm (or 440 nm) after 10 min. Set up the appropriate blanks and compare the colour equivalence of the amino acids investigated.

Experiment 8.2   Biuret assay

MATERIALS   100
1. Protein standard (5 mg albumin/ml). Prepare fresh   1 litre
2. Biuret reagent. (Dissolve 3 g of copper sulphate ($CuSO_4 \cdot 5H_2O$) and   2 litres
    9 g of sodium potassium tartrate in 500 ml of 0.2 mol/litre sodium
    hydroxide; add 5 g of potassium iodide and make up to 1 litre with
    0.2 mol/litre sodium hydroxide)
3. Water bath at 37°C   20

METHOD

Add 3 ml of biuret reagent to 2 ml of protein solution, mix, and warm at 37°C for 10 min; cool, and read the extinction at 540 nm. Prepare a graph of extinction against albumin concentration: this standard curve will prove useful in other experiments.

Experiment 8.3   The Folin–Lowry method of protein assay

PRINCIPLE

Protein reacts with the Folin–Ciocalteau reagent to give a coloured complex. The colour so formed is due to the reaction of the alkaline copper with the protein as in the biuret test and the reduction of phosphomolybdate by tyrosine and tryptophan present in the protein. The intensity of colour depends on the amount of these aromatic amino acids present and will thus vary for different proteins.

MATERIALS   100
1. Alkaline sodium carbonate solution (20 g/litre Na_2CO_3 in   3 litres
    0.1 mol/litre NaOH)
2. Copper sulphate–sodium potassium tartrate solution (5 g/litre   100 ml
    $CuSO_4 \cdot 5H_2O$ in 10 g/litre Na, K tartrate). Prepare fresh by mixing
    stock solutions
3. 'Alkaline solution'. Prepare on day of use by mixing 50 ml of (1) and   3 litres
    1 ml of (2)
4. Folin–Ciocalteau reagent. (Dilute the commercial reagent with an   500 ml
    equal volume of water on the day of use. This is a solution of sodium
    tungstate and sodium molybdate in phosphoric and hydrochloric
    acids)
5. Standard protein (albumin solution 0.2 mg/ml)   1 litre

METHOD

Add 5 ml of the 'alkaline solution' to 1 ml of the test solution. Mix thoroughly and allow to stand at room temperature for 10 min or longer. Add 0.5 ml of diluted Folin–Ciocalteau reagent *rapidly with immediate mixing.* After 30 min read the extinction against the appropriate blank at 750 nm.

Estimate the protein concentration of an unknown solution after preparing a standard curve.

Experiment 8.4   A comparison of the various methods of protein estimation

Compare the biuret and the Folin–Lowry methods for the estimation of proteins with the UV absorption at 210 nm and 280 nm as given in Exp. 7.9. Use albumin, casein, gelatin and lysozyme as the reference proteins and determine the colour stability, reproducibility and sensitivity of the methods. In the light of these findings, outline the main advantages and limitations of each method.

The isolation of proteins

A number of physical techniques are used in the isolation and separation of proteins, and the principles behind some of these have already been considered in this chapter.

Some of the experiments detailed in the chapter on separation methods can be carried out at this stage if required, since many of the examples selected involve proteins.

Experiment 8.5   The isolation of casein from milk

PRINCIPLE

Casein is the main protein found in milk and is present at a concentration of about 35 g/litre. It is actually a heterogeneous mixture of phosphorus containing proteins and not a single compound.

Most proteins show a minimum solubility at their isoelectric point and this principle is used to isolate the casein by adjusting the pH of the milk to 4.8, its isoelectric point. Casein is also insoluble in ethanol and this property is used to remove unwanted fat from the preparation.

MATERIALS	10
1. Milk	1 litre
2. Sodium acetate buffer (0.2 mol/litre, pH 4.6)	1 litre
3. Ethanol (95 per cent v/v)	500 ml
4. Ether (*Care:* highly inflammable)	500 ml
5. Thermometer to 100°C	5
6. Muslin	—
7. Buchner filter equipment and papers	5

METHOD

Place 100 ml of milk in a 500 ml beaker and warm to 40°C, also warm 100 ml of the acetate buffer and add slowly with stirring. The final pH of the mixture should be about 4.8 and this can be checked with a pH meter. Cool the suspension to room temperature then leave to stand for a further 5 min before filtering through muslin.

Wash the precipitate several times with a small volume of water then suspend it in about 30 ml of ethanol. Filter the suspension on a Buchner funnel and wash the precipitate a second time with a mixture of equal volumes of ethanol and ether. Finally, wash the precipitate on the filter paper with 50 ml of ether and suck dry. Remove the powder and spread out on a watch glass to allow evaporation of the ether.

Weigh the casein and calculate the percentage yield of the protein. How does your figure compare with the 3.5 g theoretical yield from 100 ml of milk?

Experiment 8.6   The preparation and properties of cytochrome *c*

PRINCIPLE

When hearts are homogenized with trichloracetic acid, most of the proteins are precipitated, but cytochrome *c* is unusual in this respect and remains in solution. Ammonium sulphate is added to the TCA extract, which completes the precipitation of myoglobin, haemoglobin, and other proteins. Further acidification of the solution with TCA precipitates the cytochrome *c* which is then dialysed and characterized.

MATERIALS	<u>10</u>
1. Pig or ox heart (1 kg per two students)	5 kg
2. Large mincer	2
3. Muslin	—
4. Sodium hydroxide (100 g/litre)	500 ml
5. Ammonium sulphate	—
6. Trichloracetic acid (200 g/litre)	500 ml
7. Trichloracetic acid (0.145 mol/litre)	6 litres
8. Saturated ammonium sulphate solution	1 litre
9. Dialysis tubing	—
10. pH meter	2
11. Spectrophotometer	5
12. Waring blender	5
13. Potassium ferricyanide (10 mmol/litre)	50 ml
14. Sodium dithionite saturated solution	50 ml

METHOD

Remove the fat from the hearts and weigh them. Mince 1 kg of the hearts and blend with 1 litre of 0.145 mol/litre TCA. Allow the suspension to stand at room temperature for 3 h then filter through muslin. Adjust the pH of the cloudy fluid obtained to pH 7.3 with 100 g/litre NaOH. Measure the total volume of the fluid and slowly add solid ammonium sulphate (500 g/litre) with stirring. Filter off the precipitate with a large fluted filter paper to obtain a pink filtrate. Measure the volume and add further ammonium sulphate (50 g/litre) with stirring. Leave the mixture overnight in the refrigerator.

Remove any slight precipitate formed overnight and add 200 g/litre TCA (25 ml/litre) to precipitate the cytochrome c. Rapidly centrifuge off the cytochrome c (3000 g for 15 min) and suspend the brick-red precipitate in saturated ammonium sulphate solution (150 ml/kg original tissue). Place the suspension in a dialysis sac and dialyse against water for 4 h. Centrifuge off the slight, dark brown precipitate of denatured cytochrome c and store the remaining solution at $-15°C$.

The cytochrome c can conveniently be characterized in solution or alternatively precipitated by the addition of 4 volumes of cold acetone. A better product is obtained if the solution is dialysed free of ammonium sulphate and freeze dried.

Record the final volume of the cytochrome c solution (v).

Spectrophotometric standardization The solution prepared is a mixture of the oxidized and reduced forms of the molecule. The next stage is to oxidize one sample with ferricyanide, to reduce another sample with dithionite, and to record their extinctions at 550 nm. The molar extinction coefficients are known for each form so the purity can be readily checked. If other pigments are present, it is unlikely that they will show the same shift when converted from the oxidized to the reduced form and vice versa. The molar extinction coefficients at 550 nm are:

$$\text{Oxidized form, } k_1 = 0.9 \times 10^4 \text{ (litre mol}^{-1}\text{ cm}^{-1}\text{)}$$

$$\text{Reduced form, } k_2 = 2.77 \times 10^4 \text{ (litre mol}^{-1}\text{ cm}^{-1}\text{)}$$

This means that a solution of oxidized cytochrome c of concentration 1 mol/litre or 1 mmol/ml has an extinction of 0.9×10^4 in a light path of 1 cm.

Oxidized form Prepare the following solution and read the extinction E_1 at 550 nm against a blank of ferricyanide and buffer.

Component	ml
Sodium phosphate buffer (0.1 mol/litre, pH 7.4)	
Cytochrome c preparation suitably diluted	1.0
Potassium ferricyanide (0.01 mol/litre)	0.1

Reduced form Prepare the following mixture and read the extinction E_2 against a blank of dithionite and buffer.

Component	ml
Sodium phosphate buffer (0.1 mol/litre, pH 7.4)	1.9
Cytochrome c preparation	1.0
Sodium dithionite saturated solution	0.1

Use the data obtained to calculate the concentration of cytochrome c present in the cuvette as mg/ml. Multiply this figure by three as there are 3 ml in the cuvette and remember that this amount of cytochrome c was present in 1 ml of the diluted preparation. The yield of cytochrome c in milligrams (Y) is therefore given by:

$$Y = (E_1/k_1) \times \text{mol. wt} \times 3 \times \text{dilution} \times V\,\text{mg}$$

and

$$Y = (E_2/k_2) \times \text{mol. wt} \times 3 \times \text{dilution} \times V\,\text{mg}$$

Express the final result as milligrams of cytochrome c per kilogram of heart. The yields calculated from the oxidized and reduced forms should be the same if the material is relatively pure. As a final check on purity, determine the total amount of protein present using one of the standard procedures already described.

Absorption spectra   Finally, plot the absorption spectra of the oxidized and reduced forms and determine the absorption maxima in the visible region.

Oxidized maxima: 408 nm and 530 nm.
Reduced maxima: 415 nm, 520 nm and 550 nm.

Protein structure

Experiment 8.7   The identification of the C-terminal amino acid of a protein

PRINCIPLE

The enzyme carboxypeptidase is an exopeptidase that degrades polypeptides by catalysing the hydrolysis of the peptide bond next to the C-terminal amino acid. This release of the C-terminal amino acid unmasks a new C-terminal residue which is then in turn cleaved by the enzyme. The C-terminal amino acid can thus be identified and also the sequence of several amino acids close to this.

In this experiment, carboxypeptidase A is incubated with several proteins and, during the digestion, samples are removed and the amino acids identified by paper chromatography.

MATERIALS   10
1. Tris–HCl buffer (25 mmol/litre, pH 7.5)   50 ml
2. Protein solutions in buffer (muramidase and ribonuclease, 10 mg/ml)   10 ml
3. Carboxypeptidase A (1 mg/ml in buffer)   10 ml
4. Water baths at 37°C   5
5. Micropipettes (50 µl)   10
6. Equipment for paper chromatography of amino acids (Exp. 4.4)   5
7. Trichloracetic acid (100 g/litre)   50 ml

METHOD
Add 0.5 ml of carboxypeptidase A to 0.5 ml of the protein solution, mix thoroughly, and place in a water bath at 37°C. Withdraw 0.2 ml samples at suitable time intervals

(0, 10, 20, 30 and 60 min) and mix with 0.2 ml of 10 per cent w/v TCA. Centrifuge the precipitate and spot 50 μl samples on to Whatman No. 1 chromatography paper. Separate the amino acids and identify them as far as possible (Exp. 4.4), also plot a graph of the number of amino acids appearing with time. The experiment can be shortened by separating the amino acids in one dimension only, using the butanol/acetic acid/water solvent, and setting up a limited number of standard amino acids (alanine, arginine, leucine, serine and valine).

Experiment 8.8   The determination of the free amino end group of some proteins

PRINCIPLE

The *N*-terminal amino acid can be identified with 1-fluoro-2,4-dinitrobenzene (Sanger's reagent) which reacts in mildly alkaline solution with the free amino group at the end of the chain.

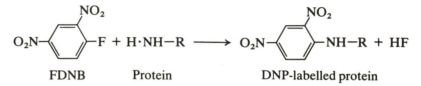

$$O_2N-\langle\rangle-F + H\cdot NH-R \longrightarrow O_2N-\langle\rangle-NH-R + HF$$

FDNB   Protein   DNP-labelled protein

The protein is hydrolysed in acid solution and since the DNP–amino acid link is resistant to this treatment, the end amino acid is effectively labelled. After hydrolysis, the yellow DNP–amino acid is identified by chromatography.

FDNB also reacts with free amino, imidazole and phenolic groups on amino acids at neutral to alkaline pH to give the corresponding DNP derivatives, but, fortunately, the non-polar DNP–amino acids can be readily extracted from the acid hydrolysate with ether, leaving these charged DNP–amino acid derivatives in the aqueous phase.

MATERIALS	100
1. Proteins (haemoglobin, muramidase and ribonuclease)	3 g
2. Sodium bicarbonate	—
3. Hydrochloric acid (conc.)	500 ml
4. Ether (peroxide free)	2 litres
5. 1-Fluoro-2,4-dinitrobenzene (5 per cent v/v in ethanol). (*Care:* causes blisters)	1 litre
6. Hydrochloric acid (6 mol/litre)	500 ml
7. Sodium phthalate buffer (0.1 mol/litre, pH 4.6)	1 litre
8. Chromatography equipment	—
9. Whatman No. 4 chromatography paper	—
10. Shakers (capacity 4 tubes)	25
11. Oven at 110°C	10
12. Acetone	100 ml
13. Sodium phosphate buffer (0.75 mol/litre, pH 6.0)	Capacity of tanks

14. Ampoules   150
15. Ultraviolet lamp   20
16. Standard DNP derivatives of glycine, valine, lysine, and   —
    phenylalanine. (Keep in the dark)

METHOD

Preparation of the DNP–amino acid   Weigh out about 5–20 mg of protein (minimum 0.2 μmol), mix with an equal weight of sodium carbonate and suspend in 1–2 ml of water. Add twice the volume of FDNB solution and shake for 2 h at room temperature. Maintain the pH in the region of 8–9 by adding more sodium bicarbonate if required. If a large precipitate forms, the pH is too low. Extract the suspension three times with peroxide-free ether to remove dinitrophenol formed from reaction of the FDNB with water. Adjust the pH to about 1 using strong acid and extract three times with an equal volume of peroxide-free ether; combine the ether extracts and evaporate to dryness in a fume chamber. Add 0.2 ml of acetone to the dried DNP derivative and transfer to a hydrolysis vial. Remove the acetone by evaporation in a stream of air and add 1 ml of 6 mol/litre HCl. Seal the ampoule and place in an oven at 110°C for 18 h.

Carefully open the vial after cooling it to room temperature, add 1 ml of water, and extract three times with about 2 ml of ether. Concentrate the combined ether extracts and evaporate to dryness. Dissolve the DNP–amino acid in a little acetone and chromatograph as below.

Chromatography of DNP–amino acids   Apply 10–20 μl of the test to Whatman No. 4 paper previously saturated with phthalate buffer and repeat this until the yellow spots are clearly visible. Apply the standard DNP–amino acid solutions and develop the ascending chromatogram with 0.75 mol/litre phosphate buffer, pH 6.0.

Experiment 8.9   The detection of changes in the conformation of bovine serum albumin by viscosity measurements

PRINCIPLE

High concentrations of urea cause an unfolding of proteins by weakening the hydrophobic bonds that maintain the tertiary structure. This change in protein conformation leads to a less compact molecule with a larger viscosity than the native protein. Such changes in tertiary structure can be readily followed using an Ostwald viscometer (Fig. 8.1). This, essentially, consists of a capillary tube down which a known volume of protein solution is allowed to flow under gravity. The time taken for this flow is measured (t_1) and also that of the solvent (t_0); the relative viscosity is then given by:

$$\eta_{rel} = \eta_1/\eta_0 = (t_1/t_0) \times (\rho_1/\rho_0)$$

where η_1 is the viscosity of the protein solution of density ρ_1 and η_0 the viscosity of the solvent of density ρ_0. If the densities are taken to be the same then the expression

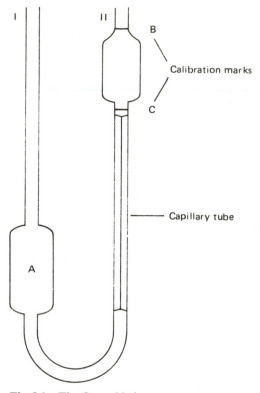

Fig. 8.1   The Ostwald viscometer

simplifies to:

$$\eta_{rel} = t_1/t_0$$

Einstein has shown that, for spherical molecules, the relative viscosity is related to the concentration of the molecule (c) and the partial specific volume ($\bar{V}$), which is the volume occupied by the molecule and its bound water:

$$\eta_{rel} = 1 + 2.5c\bar{V}$$

MATERIALS   <u>10</u>
Viscosity is very sensitive to temperature, so all solutions and the
viscometer must be kept at 30°C in the water bath.
1. Ostwald viscometer   5
2. Water bath at 30°C   5
3. Potassium chloride (100 mmol/litre)   4 litres
4. Urea solutions (0.5, 1, 2, 4, 6, and 8 mol/litre in 100 mmol/litre KCl) 250 ml
5. Bovine serum albumin (10 g/litre in 100 mmol/litre KCl and the   250 ml
   above urea solutions)
6. Stop watch accurate to at least 0.1 s   5

METHOD

Always handle the viscometer by one limb only and never squeeze the two arms together. Rinse the viscometer with KCl solution and place it in position in water bath by carefully clamping one limb. Check that it is vertical using a plumbline and introduce exactly 20 ml (or the volume marked on the viscometer) of KCl solution at 30°C into the bulb A with a syringe or pipette. Leave for 5 min to equilibrate, then either apply positive pressure to the wide limb (I) or gentle suction to the other limb (II) until the meniscus rises above the upper graduation mark B. Release the pressure and measure the time (to the nearest 0.1 s) for the liquid to flow between the two graduation marks B and C. Repeat the experiment until the flow times agree within 0.2 s and calculate the average flow time. Repeat the whole procedure with the urea solutions alone (t_0), which are the solvents, and then with the bovine serum albumin dissolved in the urea (t_1). Plot the values of t_0 and t_1 against the concentration of urea and join up the points with smooth curves. Select convenient concentrations of urea and calculate the relative viscosities (t_1/t_0) using the values from the curves. This ensures that any slight errors involved in the determination of t_1 and t_0 are not magnified on taking the ratios. Finally, prepare a graph of the relative viscosity against the concentration of urea and comment on the results. In addition, calculate the partial specific volume of serum albumin in 10 mmol/litre KCl and in 8 mol/litre urea. Assume that the molecule remains spherical so that Einstein's equation is valid.

Experiment 8.10   The effect of pH on the conformation of bovine serum albumin

MATERIALS

1. As for Exp. 8.9

2. pH meter

$$\frac{10}{5}$$

METHOD

Using the Ostwald viscometer, follow the structural changes in the albumin dissolved in 100 mmol/KCl and distilled water as the pH is varied over the range 2–12. Comment on the results.

Further reading

Alexander, P. and Lundgren, H. P., *A Laboratory Manual of Analytical Methods of Protein Chemistry*, Vols 1–5. Pergamon Press, Oxford, 1966–1969.

Creighton, T. E., *Proteins: Structure and Molecular Principles*. W. H. Freeman, New York, 1983.

Hughes, R. C., *Glycoproteins*. Chapman and Hall, London, 1983.

Kerese, I., *Methods of Protein Analysis*. Wiley, New York, 1984.

Scopes, R., *Protein Purification: Principles and Practice*. Springer-Verlag, New York, 1982.

Van Holde, K. E., *Physical Biochemistry*, 2nd edn. Prentice-Hall, Englewood Cliffs, 1985.

9. Carbohydrates

The function of carbohydrates in the biosphere

Carbohydrates are carbon compounds that contain hydrogen and oxygen in the ratio of 2 to 1. The term is also used for the derivatives of carbohydrates where the above definition may not be strictly true.

$$C_x(H_2O)_y$$

$$\uparrow \quad \uparrow$$

Carbon   Hydrate

Carbohydrate

A source of energy

These compounds are of fundamental importance in living organisms as a source of metabolic energy.

Photosynthesis They are synthesized in green plants and algae from water and CO_2 using the energy of sunlight—a process known as photosynthesis. Many plants thus contain large quantities of carbohydrates as food reserves which are then eaten by man and other animals. After ingestion, the plant carbohydrates are broken down to glucose which is stored as glycogen in liver, muscle and other tissues.

Respiration In the animal, the stored carbohydrate is converted to glucose which is then transported to the cells where it is oxidized to CO_2 and O_2 in cellular respiration. The energy released during this oxidation is then 'captured' and used to drive the metabolic machinery of the cells. In the case of man, more than 60 per cent of his total energy requirements are provided by the oxidation of carbohydrates. In this way the energy of sunlight is made available to the animal kingdom which cannot carry out photosynthesis.

The overall process is summarized in Fig. 9.1.

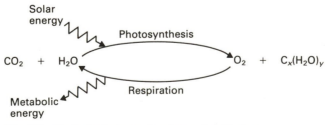

Fig. 9.1 Carbohydrate production and oxidation

Structure of cells and molecules

Carbohydrates are also important components of some of the structural materials of living organisms. Some examples of this are the cell walls in plants, polysaccharides in the capsules of bacteria, and mucopolysaccharides of skin and connective tissue in animals. In addition, monosaccharides are an important part of biochemical compounds such as nucleic acids, coenzymes and flavoproteins. Carbohydrates are also involved in cell recognition, contact inhibition and the antigenic properties of blood group substances.

The structure of carbohydrates

Introduction

The basic units of carbohydrates are *monosaccharides* which cannot be split further by hydrolysis. Chemically they are polyhydroxyaldehydes (aldoses) or polyhydroxyketones (ketoses). They are named according to the number of carbon atoms in the chain so that *tetroses* contain four, *pentoses* five and *hexoses* six carbon atoms.

Stereochemistry

Optical activity   Many biological molecules, including sugars, contain one or more *asymmetric carbon atoms* or *chiral centres* and, because of this, a number of stereoisomers are possible. For example, glyceraldehyde, one of the simplest sugars, has one asymmetric carbon atom and there are two possible arrangements of the four groups around this carbon. The formulae of these two isomers are given below, where the asymmetric carbon is placed at the centre of a tetrahedron, the four groups being situated at each corner, and the dotted line shown as lying below the plane of the paper. The two forms can be likened to mirror images or left- and right-handed gloves.

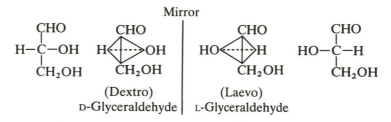

Mirror

(Dextro)   (Laevo)
D-Glyceraldehyde | L-Glyceraldehyde

These two configurations are *optical isomers* or *enantiomers* and rotate the plane of polarized light to the same extent but in opposite directions. When the plane of polarized light is rotated to the right, the compound is *dextrorotatory* and is labelled (*d*) or (+). Likewise, when the plane of polarized light is rotated to the left, the compound is *laevorotatory* (*l*) or (−).

The D *and* L *forms*   Aldoses are derived from the parent compound, glyceraldehyde, by the addition of successive secondary alcohol groups (—CHOH). Since glyceraldehyde can exist in two forms, two distinct families of aldoses emerge: those derived from *d*-glyceraldehyde known as D-*sugars* and those from *l*-glyceraldehyde called L-*sugars*. The letters D and L do not give an indication of the optical activity, but refer to the configuration of the carbon atom next but one furthest removed from the end of the chain containing the aldehyde or ketone group. Thus, a D sugar could be dextrorotatory (D(+)) or laevorotatory (D(−)) depending on the configuration of the other carbon atoms present.

$$\text{Aldoses} \quad \begin{array}{c} \text{CHO} \\ | \\ \text{(CHOH)}_n \\ | \\ \text{CH}_2\text{OH} \end{array}$$

		Number of forms	
		D	L
Trioses	$n = 1$	1	1
Tetroses	$n = 2$	2	2
Pentoses	$n = 3$	4	4
Hexoses	$n = 4$	8	8

The simplest ketose is the triose, dihydroxyacetone, but this compound is optically inactive so the tetrose D-erythrulose is the parent compound of the D-ketoses.

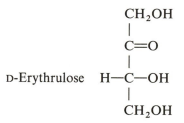

D-Erythrulose
$$\begin{array}{c} \text{CH}_2\text{OH} \\ | \\ \text{C}{=}\text{O} \\ | \\ \text{H}-\text{C}-\text{OH} \\ | \\ \text{CH}_2\text{OH} \end{array}$$

Pentoses, hexoses, etc., are formed by the addition of secondary alcohol groups.

		Number of forms	
		D	L
Tetroses	$n = 1$	1	1
Pentoses	$n = 2$	2	2
Hexoses	$n = 3$	4	4

Numbering of carbon atoms   The carbon atoms are numbered from the end of the chain containing the reactive carbonyl group thus:

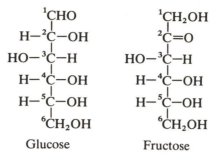

Glucose   Fructose

Ring formulae   When D(+)-glucose is dissolved in water, a specific rotation of +113 degrees is obtained, but this slowly changes, so that at 24 h the value has become +52.5 degrees. This phenomenon is known as *mutarotation* and is shown by a number of pentoses, hexoses and reducing disaccharides. The reason for this change in rotation is that glucose exists in solution mainly in the ring form, which is obtained by the formation of an intramolecular hemiacetal, as below.

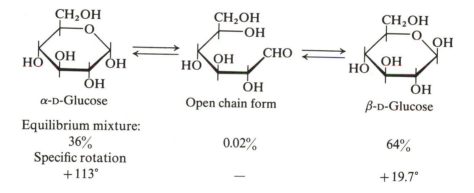

α-D-Glucose   Open chain form   β-D-Glucose

Equilibrium mixture:
36%   0.02%   64%
Specific rotation
+113°   —   +19.7°

This creates another asymmetric carbon atom (C_1) so that two ring forms (α and β) are now possible and these are known as *anomers*.

The open chain form is in equilibrium with the two ring forms and this is responsible for the reducing properties of the sugar. The above ring formula shown for glucose is that proposed by *Haworth* and shows the ring as a plane with the hydroxyl groups orientated above and below the plane. For the sake of clarity the hydrogen atoms are not usually shown.

The ring formed in the case of glucose contains five carbon atoms and one oxygen and is analogous to pyran; glucose is said, therefore, to exist in the *pyranose* form. Many of the common sugars are present as the pyranose ring, but some exist as a four-carbon ring known as the *furanose* form, after the compound furan.

Pyran   Furan

The structures of a number of common monosaccharides are given in Fig. 9.2.

The glycosidic link

Glycosides and disaccharides   The carbonyl group present in all sugars is very reactive and can form hemiacetals or acetals with other hydroxylic compounds.

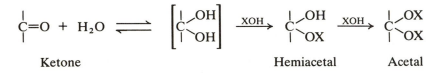

Ketone   Hemiacetal   Acetal

Formation of an internal ring structure is brought about by reaction of the hydroxyl group on C_4 or C_5 with the carbonyl group to give an intramolecular hemiacetal,

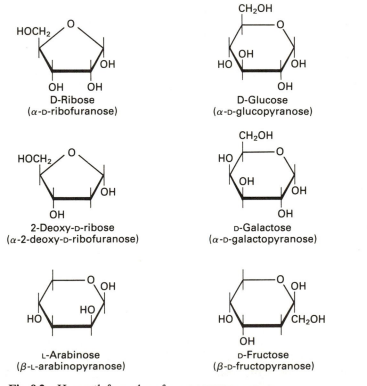

D-Ribose
(α-D-ribofuranose)

D-Glucose
(α-D-glucopyranose)

2-Deoxy-D-ribose
(α-2-deoxy-D-ribofuranose)

D-Galactose
(α-D-galactopyranose)

L-Arabinose
(β-L-arabinopyranose)

D-Fructose
(β-D-fructopyranose)

Fig. 9.2   Haworth formulae of some common sugars

where XOH is the rest of the molecule. The remaining hydroxyl on C_1, known as the *glycosidic hydroxyl*, is very reactive and readily forms a glycosidic link with other hydroxyl groups by the elimination of water. If the hydroxylic compound is not a sugar, then a *glycoside* is formed, as, for example, in the condensation of glucose and methanol to give methyl glucoside. Quite often, though, the hydroxyl is from another sugar and in this case a *disaccharide* is the product. Providing the bond is not between two C_1 atoms, then a free aldehyde or ketone group is left on the disaccharide and the molecule shows all the reactions associated with these groups as well as mutarotation.

Oligosaccharides and polysaccharides   This process can be repeated so that a second monosaccharide is linked to a third by another glycosidic bond to give a *trisaccharide* and so on to give an *oligosaccharide* of 2–10 units linked as a chain. In the case of *polysaccharides*, chains of 10 to several thousand monosaccharide units are joined together by glycosidic bonds to give a very large molecule. The monomer units are not always the same and branching can occur as well as substitution to give a quite complex molecule.

Simple monosaccharide derivatives

Oxidized products   The aldehyde and primary alcohol groups of aldoses can be oxidized to the corresponding *aldonic* and *uronic* acids. Further oxidation of these acids yields *saccharic* acid. The oxidation of a monosaccharide is shown below. The names in brackets represent the products when the hexose is the monosaccharide glucose.

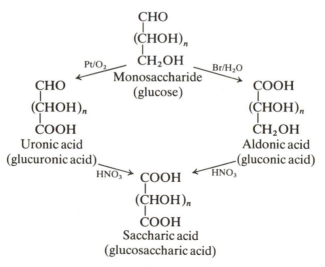

Reduction   The aldehyde or ketone group present can be reduced to the primary or secondary alcohol with sodium amalgam. For example, fructose and glucose give the hexahydric alcohol *sorbitol* and glyceraldehyde is reduced to *glycerol*.

$$
\begin{array}{ccc}
\text{CHO} & & \text{CH}_2\text{OH} \\
| & & | \\
\text{CHOH} & \xrightarrow{\ 2\text{H}\ } & \text{CHOH} \\
| & & | \\
\text{CH}_2\text{OH} & & \text{CH}_2\text{OH} \\
\text{Glyceraldehyde} & & \text{Glycerol}
\end{array}
$$

Esters   Sugars form esters very readily with acids by reaction with the acid chloride or anhydride in the presence of a catalyst. In particular, the phosphate esters of the monosaccharides are of fundamental importance in carbohydrate metabolism. Esters formed from the primary alcohol group of the ultimate carbon (glucose-6-phosphate) are usually the most stable to acid hydrolysis. Hemiacetals formed through the reducing group are, on the other hand, more acid-labile (glucose-1-phosphate).

Amino derivatives   The hydroxyl in the 2 position of many sugars can be replaced by an amino group, as for example in *galactosamine* and *glucosamine*. These compounds also exist as the *N*-acetyl derivatives which are part of the structure of several polysaccharides. The presence of the amino group close to the glycosidic bond renders it more stable to hydrolysis.

Table 9.1   Some common mono- and disaccharides

Sugar	Occurrence and function
Monosaccharides	
Pentoses	
L-Arabinose	Present as pentosans in wood gums and fermented by some bacteria. The D sugar occurs in the glycoside of tubercle bacilli.
D-Ribose	An essential part of RNA the macromolecule involved in protein synthesis, also present in many coenzymes (NAD, FAD, ATP).
2-Deoxy-D-ribose	An important constituent of the macromolecule DNA, the genetic material of all living organisms.
Hexoses	
D-Glucose	The most widely distributed of the sugars; glucose is transported in the blood and oxidized in the cells to produce energy.
D-Fructose	The sweetest of the sugars—found in fruit and honey. It is also present in seminal fluid where it provides the energy source for the spermatozoa.
D-Galactose	This sugar is part of the glycolipids of nervous tissue and chloroplast membranes. Galactosamine is present in blood group substances, cartilage and tendons.
Disaccharides	
Lactose (β-D-galactosyl-1,4-D-glucose)	The sugar present in the milk of mammals which may also be found in the urine during pregnancy.
Maltose (α-D-glucosyl-1,4-D-glucose)	The sugar produced during the digestion of starch by amylase and present in germinating cereals and malt.
Sucrose (α-D-glucosyl-β-1,2-D-fructose)	This is a non-reducing sugar since the glycoside link is through both of the potentially reducing groups of the monosaccharides. Occurs in plants and is obtained mainly from sugar cane and beet.

Carbohydrates of biochemical importance

Simple sugars

Simple. sugars are widespread throughout nature and perform many and varied functions. The list in Table 9.1 is by no means comprehensive and, for the sake of simplicity, carbohydrate derivatives have not been included.

Macromolecules

Homopolysaccharides These are very large molecules which contain only one monosaccharide component, although several macromolecules can be built up from the same monosaccharide unit. This multiplicity arises from the different size of the molecules, the variable degree of branching and cross-linking, and the nature of the glycosidic bond. It is illustrated nicely in the case of glucose, which is the basic component of several polysaccharides (Table 9.2).

Heteropolysaccharides These molecules are formed from two or more basic units and are frequently associated with protein. Such complexes are known as *proteoglycans* or *mucoproteins* when the polysaccharide component dominates. These

Table 9.2 The properties of some homopolysaccharides

Polysaccharide (*monosaccharide unit*)	*Straight chain*	*Branch points*	*Mol. wt*
Storage polysaccharides			
Amylopectin (α-D-glucose)	α 1–4	α 1–6	$0.2–1 \times 10^6$
The major component of starch the main carbohydrate store in plants.			
Amylose (α-D-glucose)	α 1–4	none	50 000
Present in most starches making up to 20 per cent of the total.			
Glycogen (α-D-glucose)	α 1–4	α 1–6	$1–3 \times 10^6$
The major reserve carbohydrate in animals present mainly in the liver and muscle.			
Inulin (β-D-fructose)	β 1–2	none	5000
The reserve carbohydrate of plants such as dahlias and artichokes.			
Structural polysaccharides			
Cellulose (β-D-glucose)	β 1–4	none	—
The major structural component of the cell walls of plants and some algae and bacteria.			
Chitin (β-D-*N*-acetyl-glucosamine)	β 1–4	none	—
The main component of the exoskeleton of insects and crustacea such as crabs and lobsters.			

materials are gelatinous and viscous in nature and frequently act as lubricants or intracellular cement. The carbohydrate–protein complexes are called *glycoproteins* when the carbohydrate is the minor component. The small quantities of short-chain oligosaccharides can have a considerable influence on the biological properties of the proteins. For example, the antigenic properties of blood group substances depend on the chemical nature of these carbohydrate chains. The structure and linkages are not always known for certainty so only the components are given in Table 9.3.

Table 9.3   Some common heteropolysaccharides

Polysaccharide	Component monosaccharides
Mucopolysaccharides	
Heparin	D-Glucuronic acid and D-glucosamine sulphate
This is present in the arterial walls and lungs as a natural anticoagulant. It is also used as such in the laboratory and in certain cases therapeutically.	
Hyaluronic acid	D-Glucuronic acid and *N*-acetyl-D-glucosamine
A component of the extracellular cement; also present in the synovial fluid round joints where it acts as a lubricant.	
Chondroitin sulphates	D-Glucuronic acid and *N*-acetyl-D-galactosamine sulphate
Very abundant in skin and cartilaginous tissue.	
Bacterial polysaccharides	
Murein	*N*-acetyl-D-glucosamine, *N*-acetylmuramic acid and oligopeptide chains
A large three-dimensional network forming the backbone of bacterial cell walls.	
Capsules	Exact composition depends on the organism
This polysaccharide envelope is responsible for certain of the antigenic properties of the bacteria.	

Chemical properties

Experiment 9.1   Benedict's test for reducing sugars

PRINCIPLE

If a suspension of copper hydroxide in alkaline solution is heated, then black cupric oxide is formed:

$$Cu(OH)_2 \rightarrow CuO + H_2O$$

However, if a reducing substance is present, then rust-brown cuprous oxide is precipitated:

$$2Cu(OH)_2 \rightarrow Cu_2O + 2H_2O + \tfrac{1}{2}O_2$$

In practice, an alkaline solution of a copper salt and an organic compound containing alcoholic —OH is used rather than the above suspension. Under these conditions, the copper forms a soluble complex and the reagent is stable.

Carbohydrates with a free or potentially free aldehyde or ketone group have reducing properties in alkaline solution. In addition, monosaccharides act as reducing agents in weakly acid solution.

Benedict modified the original Fehling's test to produce a single solution which is more convenient for tests, as well as being more stable, than Fehling's reagent.

MATERIALS   100
1. Benedict's reagent. (Dissolve 173 g of sodium citrate and 100 g   3 litres
   sodium carbonate in about 800 ml of warm water. Filter through a
   fluted filter paper into a 1000 ml measuring cylinder and make up to
   850 ml with water. Meanwhile, dissolve 17.3 g of copper sulphate in
   about 100 ml of water and make up to 150 ml. Pour the first solution
   into a 2 litre beaker and slowly add the copper sulphate solution
   with stirring)
2. Glucose solutions (10 g/litre and 1 g/litre)   1 litre

METHOD
Add five drops of the test solution to 2 ml of Benedict's reagent and place in a boiling water bath for 5 min. Examine the sensitivity of Benedict's test using increasing dilutions of glucose.

Experiment 9.2   Iodine test for polysaccharides

PRINCIPLE
Iodine forms coloured adsorption complexes with polysaccharides; starch gives a blue colour with iodine while glycogen and partially hydrolysed starch react to form red–brown colours.

MATERIALS   100
1. Iodine solution (5 mmol/litre in KI (30 g/litre))   100 ml
2. Cellulose, glycogen, starch and inulin (10 g/litre)   200 ml

METHOD
Acidify the test solution with dilute HCl, then add two drops of iodine, and compare the colours obtained with that of water and iodine.

Optical activity

THE POLARIMETER
As previously discussed, most sugars contain one or more asymmetric carbon atoms and show optical activity. This rotation of the plane of polarized light can be demonstrated and measured with a polarimeter (Fig. 9.3). Monochromatic light passes through a *Nicol prism* and emerges polarized in one plane. This polarized beam then passes through the sugar sample which rotates the plane of the light. The second *Nicol prism* is rotated until its plane of polarization lies at right angles to that of the first prism and the light beam is prevented from passing through the instrument.

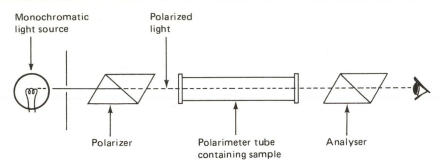

Fig. 9.3   The essential features of a polarimeter

Alternatively, the second prism is rotated so that it corresponds with the plane of polarization produced by the first prism to give a field of maximum brightness. The instrument is zeroed on either of these positions with only solvent in the sample chamber. In practice, the emergent light is seen as two semicircular zones and the zero is obtained when these two halves of the field of view are of uniform darkness or brightness. The solvent is then replaced with the solution to be investigated and the analyser rotated to restore the situation of minimum or maximum brightness. The degree and direction of the rotation are then recorded.

   If the angle of rotation is *clockwise* then the compound is *dextrorotatory* $(+)$ and if *anticlockwise* then the sugar is *laevorotatory* $(-)$.

SPECIFIC ROTATION

The degree of rotation recorded depends on a number of factors including the length of the light path (l dm) and the concentration of the solute (c g/ml) as well as the temperature ($t°C$) and the wavelength of the light used (λ nm). The *specific rotation* is characteristic of a particular compound and is defined as the rotation of monochromatic light caused by 1 g/ml of optically active solute in a 1 dm tube at a fixed temperature.

$$\text{Specific rotation } [\alpha]_\lambda^t = \alpha/(l \times c)$$

   In practice, most simple polarimeters use sodium lamps ($\lambda = 589$ nm) and observations are generally made at 20°C and, in this case, the specific rotation is presented as $[\alpha]_D^{20}$.

Experiment 9.3   The mutarotation of glucose

PRINCIPLE

D-Glucose can be crystallized in either the α or β form, and freshly prepared solutions of these anomers have specific rotations $[\alpha]_D^{20}$ of $+113°$ and $+19°$ respectively. On standing, these solutions show mutarotation and an equilibrium mixture of the α and β forms is obtained with a specific rotation of $+52.5°$.

MATERIALS

<u>10</u>

1. Polarimeter — 5
2. α-D-Glucose — 100 g
3. β-D-Glucose — 100 g
4. Sodium carbonate (0.1 mol/litre) — 50 ml
5. Stop clocks — 5

METHOD

Mutarotation in distilled water   Rinse the polarimeter tube with distilled water and fill completely with water. Add the last few drops with a Pasteur pipette and screw on the cap carefully to ensure that no air bubbles are trapped. Adjust the instrument to zero with the tube filled with water in the polarimeter. Empty the polarimeter tube and thoroughly dry it. Carefully transfer 5 g of α-D-glucose to a dry 50 ml flask, add 40 ml of water to dissolve the glucose, and make up to the mark with distilled water. Rapidly mix the solution and fill the polarimeter tube with the glucose: obtain a reading for the rotation as soon as possible and start the stop clock (zero time). Measure the rotation every 10 min for the next 30 min and then at longer time intervals until a constant value is obtained. Repeat the experiment with β-D-glucose and plot a graph of the change in specific rotation with time.

Mutarotation in alkali   Repeat the above experiment with the α and β forms of D-glucose but this time add 1 ml of Na_2CO_3 solution (0.1 mol/litre) before making up to the mark. Obtain the first reading as soon as possible and thereafter take readings every 2 min for the first 10 min then at longer time intervals as appropriate until no further change is observed. Plot a graph of the change in specific rotation with time.

How do your readings for specific rotation compare with those discussed above?

Repeat the experiment in alkali, but this time use sucrose instead of glucose and explain the result.

Quantitative determination of carbohydrates

Experiment 9.4   Estimation of carbohydrates by the anthrone method

PRINCIPLE

The anthrone reaction is the basis of a rapid and convenient method for the determination of hexoses, aldopentoses and hexuronic acids, either free or present in polysaccharides. The blue–green solution shows an absorption maximum at 620 nm, although some carbohydrates may give other colours. The reaction is not suitable when proteins containing a large amount of tryptophan are present, since a red colour is obtained under these conditions.

The extinction depends on the compound investigated, but is constant for a particular molecule.

MATERIALS   100
1. Anthrone reagent (2 g/litre in conc. H_2SO_4)   5 litres
2. Glucose (0.1 g/litre)   2 litres
3. Glycogen (0.1 g/litre)   2 litres
4. Other carbohydrates of the same concentration if desired   2 litres

METHOD
Add 4 ml of the anthrone reagent to 1 ml of a protein-free carbohydrate solution and
rapidly mix. (*Care:* strong acid.) Place the tubes in a boiling water bath for 10 min with
a marble on top to prevent loss of water by evaporation, cool and read the extinction
at 620 nm against a reagent blank.

Prepare standard curves for the glucose and glycogen solutions and compare them.
Remember that glucose exists as the glycoside form ($C_6H_{10}O_5$) in glycogen of mol. wt
162, *not* 180. Examine the purity of a number of samples of commercial glycogen.

Experiment 9.5   Determination of reducing sugars using 3,5-dinitrosalicylic acid

PRINCIPLE
Several reagents have been employed which assay sugars by using their reducing
properties. One such compound is 3,5-dinitrosalicylic acid (DNS) which in alkaline
solution is reduced to 3-amino-5-nitrosalicylic acid.

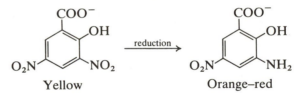

Yellow   Orange–red

The chemistry of the reaction is complicated since standard curves do not always go
through the origin and different sugars give different colour yields. The method is
therefore not suitable for the determination of a complex mixture of reducing sugars.

MATERIALS   100
1. Sodium potassium tartrate. (Dissolve 300 g of this salt in about   500 ml
   500 ml of water)
2. 3,5-Dinitrosalicylic acid. (Dissolve 10 g of this reagent in 200 ml of   200 ml
   2 mol/litre sodium hydroxide)
3. Dinitrosalicylic acid reagent. (Prepare this fresh by mixing solutions   1 litre
   (1) and (2) and making up to 1 litre with water)
4. Sodium hydroxide (2 mol/litre)   1 litre
5. Stock sugar standards (glucose, fructose and maltose 1 g/litre   50 ml
   solutions in saturated benzoic acid)
6. Working sugar standards (glucose, fructose and maltose stock   100 ml
   solutions diluted 1 in 4 before use to give solutions containing
   250 µg/ml)

7. Some sugar solutions of 'unknown' concentration   100 ml
8. Boiling water baths   50
9. Marbles   —

METHOD

Prepare the DNS reagent just before use by mixing the stock solutions as indicated and add 1 ml of the reagent to 3 ml of the sugar solution in a test tube. Prepare a blank by adding 1 ml of the reagent to 3 ml of distilled water. Cover each tube with a marble and place in a boiling water bath for 5 min, cool to room temperature and read the extinction at 540 nm against the blank. Please note that all the tubes must be cooled to room temperature before reading since the absorbance is sensitive to temperature.

Prepare standard curves of the sugars provided and use them to estimate the concentration of the 'unknowns' provided. Comment on the results.

Experiment 9.6   The determination of glucose by means of the enzyme glucose oxidase (β-D-glucose; oxygen oxidoreductase, 1.1.3.4)

PRINCIPLE

Glucose oxidase is an enzyme found in the growth medium of *Penicillium notatum* and catalyses the oxidation of β-D-glucopyranose to D-glucono-1,5-lactone with the formation of hydrogen peroxide; the lactone is then slowly hydrolysed to D-gluconic acid. The enzyme is specific for β-D-glucopyranose, but most enzyme preparations contain mutarotatase, which catalyses the interconversion of the α and β forms. D-Mannose and D-xylose are hydrolysed by the enzyme at about one-hundredth of the rate of D-glucose and the only other common sugar affected is D-galactose, which is hydrolysed at only one-thousandth of the rate. The method is therefore highly specific for glucose. Peroxidase is incorporated into the reaction mixture and catalyses the reaction of hydrogen peroxide with the chromogen ABTS (2,2'-azino-di-[3-ethylbenzthiazoline]-6-sulphonate) to give a colour which is read at 437 nm.

MATERIALS   10
1. Zinc sulphate (10 g/100 ml $ZnSO_4 \cdot 7H_2O$)   100 ml
2. Isotonic sodium sulphate (93 mmol/litre)   1 litre
3. Sodium sulphate–zinc sulphate reagent. (Dilute 55 ml of the zinc sulphate solution to 1 litre with the sodium sulphate solution)   1 litre
4. Sodium hydroxide (0.5 mol/litre)   100 ml
5. Sodium phosphate buffer (0.05 mol/litre, pH 7.0)   500 ml
6. Glucose oxidase   25 mg
7. ABTS   200 mg
8. Peroxidase   2 mg
9. Glucose oxidase reagent. Prepare this reagent fresh by dissolving 25 mg of glucose oxidase and 190 mg of ABTS in the sodium phosphate buffer. Add a small quantity of peroxidase (2 mg) and make up to 250 ml with the buffer. This solution is active for about 4 weeks if stored in a brown bottle at 4°C   —

10. Glucose standard (0.5 mmol/litre, freshly prepared)   50 ml
11. Water baths at 37°C   3

METHOD

Test   Pipette 0.1 ml of the test solution (blood, etc.) into 1.8 ml of the sodium sulphate–zinc sulphate reagent in a centrifuge tube. Add 0.1 ml of 0.5 mol/litre sodium hydroxide, centrifuge and take 0.5 ml of the supernatant in duplicate.

Blank   Take 0.5 ml of distilled water.

Standards   Use 0.5 ml of a range of glucose solutions suitably diluted from the standard.

   Add 5 ml of the glucose oxidase reagent, incubate for 1 h at 37°C and read the extinction at 437 nm against the reagent blank.

Blood sugar   Use the glucose oxidase reagent to determine the glucose concentration in blood obtained from a finger prick. Establish a 'normal range' by comparing the values obtained for the class and comment on the variation observed.

Sugars in honey   Use the DNS and the glucose oxidase methods to estimate the glucose and fructose content in the mixture of these two sugars as found in honey.

$$\beta\text{-D-Glucopyranose} + FAD \rightleftharpoons \text{D-Glucono-1,5-lactone} + FADH_2$$

$$\text{D-Glucono-1,5-lactone} + H_2O \longrightarrow \text{D-Gluconic acid}$$

$$FADH_2 + O_2 \rightleftharpoons H_2O_2 + FAD$$

$$\beta\text{-D-Glucopyranose} + H_2O + O_2 \longrightarrow \text{D-Gluconic acid} + H_2O_2$$

Experiments with polysaccharides

Experiment 9.7   The isolation and assay of glycogen from the liver and skeletal muscle of rats

PRINCIPLE

Liver glycogen   The liver glycogen maintains the level of the blood glucose and represents a central reserve of fuel for the body tissues. In a well-fed animal, glucose is converted to glycogen in the liver (*glycogenesis*) but during starvation the liver glycogen is broken down to glucose (*glycogenolysis*) and becomes depleted in about 24–48 h. After this time, the blood glucose is maintained by the synthesis of glucose from non-carbohydrate sources (*gluconeogenesis*) (Fig. 9.4).

Muscle glycogen   Muscle glycogen differs from liver glycogen in that it is not particularly affected by the state of the diet and, in the absence of violent exercise,

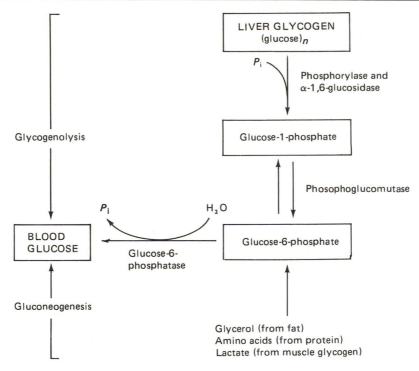

Fig. 9.4   The maintenance of blood glucose

remains fairly constant. The muscle glycogen cannot contribute directly to the blood glucose as the enzyme glucose-6-phosphatase is missing from muscle, but can do so indirectly when lactate is formed during anaerobic contractions (Fig. 9.4).

Isolation of glycogen   Glycogen is released from the tissue by heating with strong alkali and precipitated by the addition of ethanol. Sodium sulphate is added as a coprecipitant to give a quantitative yield of glycogen.

The polysaccharide is then hydrolysed in acid and the glucose released estimated.

MATERIALS   12
1. Rats of the same age, sex and weight (three fed and three starved for   6
     48 h)
2. Potassium hydroxide (300 g/litre)   100 ml
3. Calibrated centrifuge tubes   15
4. Boiling water baths   6
5. Saturated Na_2SO_4 10 ml
6. Ethanol (95 per cent v/v)   100 ml
7. Volumetric flasks (100 ml)   12
8. Test tubes calibrated at 10 ml   12

9. HCl (1.2 mol/litre)	50 ml
10. Marbles	—
11. Phenol red indicator solution	20 ml
12. NaOH (0.5 mol/litre)	50 ml
13. Reagents for the estimation of glucose (Exp. 9.6)	—

METHOD

Each pair of students should work on the liver and muscle of a fed or a starved rat.

Isolation of glycogen Kill the rat and accurately weigh about 1.5 g of liver and skeletal muscle. Place the tissues into a calibrated centrifuge tube containing 2 ml of KOH (300 g/litre) and heat in a boiling water bath for 20 min with occasional shaking. Cool the tubes in ice, add 0.2 ml of saturated Na_2SO_4, and mix thoroughly. Precipitate the glycogen by adding 5 ml of ethanol (95 per cent v/v), stand on ice for 5 min, and remove the precipitate by centrifugation. Discard the supernatant and dissolve the precipitated glycogen in about 5 ml of water with gentle warming, then dilute with distilled water to the 10 ml calibration mark and mix thoroughly. In the case of the fed animals, transfer the liver sample quantitatively to a 100 ml volumetric flask and make up to the mark with water.

Hydrolysis and estimation of glycogen Pipette duplicate 1 ml samples of the glycogen solutions into test tubes calibrated at 10 ml, add 1 ml of HCl (1.2 mol/litre), place a marble on top of each tube, and heat in a boiling water bath for 2 h. At the end of this period, add 1 drop of phenol red indicator and neutralize carefully with NaOH (0.5 mol/litre) until the indicator changes from pink through orange to a yellow colour. Dilute to 5 ml with distilled water and determine the glucose content by the glucose oxidase method (Exp. 9.6). If necessary, the samples can be stored frozen at this stage and the glucose estimation carried out at the next practical session.

Calculate the grams of glycogen per 100 g of tissue and remember that glucose exists as the glycoside form ($C_6H_{10}O_5$) in glycogen with mol. wt 162, not 180.

Prepare a table of the results obtained by other groups in the class and comment on the variation obtained and the difference between the fed and the starved animals.

Experiment 9.8 The acid hydrolysis of polysaccharides

PRINCIPLE

Acid hydrolysis of the α 1–4 and α 1–6 glycosidic bonds joining the glucose residues in glycogen is quite random, so that a whole range of oligosaccharides may be formed as intermediates although the final product is glucose.

Other polysaccharides are more resistant or more readily hydrolysed than glycogen and a few are included in this experiment for investigation.

MATERIALS	100
1. Glycogen	3 g
2. Other polysaccharides such as cellulose, inulin and chitin	3 g
3. Hydrochloric acid (2 mol/litre)	2 litres

4. Sodium hydroxide (4 mol/litre)   2 litres
5. Weak anion exchange resin   —
6. Dinitrosalicylic acid reagent (Exp. 9.5)   1 litre
7. Iodine solution (approx. 5 mol/litre) in KI (30 g/litre)   —

METHOD

Dissolve about 50 mg of glycogen in 10 ml of water and pipette 1 ml into a series of seven test tubes. Add 1 ml of water and 2 ml of 2 mol/litre HCl to each tube and place in a boiling water bath. Place a marble on top of each tube to prevent excessive loss due to evaporation and remove a tube at convenient time intervals up to 1 h. Remove one drop of the solution with a Pasteur pipette and mix with a drop of iodine on a white tile. Neutralize the contents of each tube with 1 ml of 4 mol/litre NaOH and estimate the total reducing power with the DNS reagent as set out in Exp. 9.5. Plot the progress of the hydrolysis and compare with the colour changes seen with iodine with time.

Repeat the hydrolysis with one other polysaccharide and compare with glycogen. At the end of the day obtain the results of other groups and compare the relative ease of hydrolysis of the different polysaccharides.

Meanwhile add 2 ml of the glycogen solution to 2 ml of 2 mol/litre HCl and hydrolyse for 1 h as above. Cool, then stir with several batches of the weak ion exchange resin in the hydroxyl form until the mixture has an approximately neutral pH. Decant the supernatant and use it for chromatography later. Repeat the hydrolysis with the other polysaccharides using the appropriate optimum conditions.

Experiment 9.9   Enzymic hydrolysis of glycogen by α- and β-amylase

PRINCIPLE

α-Amylase catalyses the specific hydrolysis of the α 1–4 glycosidic bonds in glycogen but is without effect on the β linkages in cellulose. The α 1–6 bonds are not affected, neither are the α 1–4 bonds linking the glucose units of maltose. Hydrolysis proceeds in a random manner with the formation of a series of intermediates. The final product is mainly maltose with some glucose and maltotriose. The increase in the number of reducing end groups as the hydrolysis proceeds is followed using the dinitrosalicylate reagent and the final products identified chromatographically.

β-Amylase from higher plants catalyses the hydrolysis of the α 1–4 bonds to give maltose, but is an exoamylase and therefore splits off maltose units from the free non-reducing end of the chain. When a branch point is reached, the digestion ceases leaving a limit dextrin. The name β-amylase is given to the enzyme as the initial product is β-maltose which later forms an equilibrium mixture of α- and β-maltose. As with α-amylase, the progress and extent of the hydrolysis are followed by determining the amount of reducing sugar produced.

MATERIALS   100
1. α-Amylase from saliva and β-amylase from plants   —
2. Rat liver glycogen preparation   1 g

3. Phosphate buffer (0.1 mol/litre, pH 6.7) containing 0.05 mol/litre   1 litre
   sodium chloride
4. Dinitrosalicylate reagent (Exp. 9.5)   2 litres
5. Water bath at 37°C   25

METHOD

Dissolve 50 mg of glycogen in 10 ml of buffered saline and pipette 1 ml into a series of seven test tubes. Add 3 ml of saliva diluted about 1 in 100 with water and incubate the tubes at 37°C for up to 1 h. Remove a tube at suitable time intervals, add 1 ml of water and stop the reaction by the addition of 1 ml of dinitrosalicylate reagent.

Meanwhile add 6 ml of α-amylase to 2 ml of glycogen solution and incubate for 1 h at 37°C. Desalt the sample with a mixed bed resin and use the supernatant for chromatography.

Repeat the above experiment with a suitably diluted preparation of β-amylase in place of the saliva and determine the amount of reducing sugar released. Compare the amount of maltose released with that obtained when α-amylase is used and calculate the percentage degradation of the glycogen by the β-amylase (α-1,4-glucan maltohydrolase, 3.2.1.2).

Experiment 9.10   The breakdown of glycogen and the production of glucose-1-phosphate by muscle phosphorylase

PRINCIPLE

Glycogen is present in large quantities in the liver and muscles of a well-fed animal. In between meals, this store of carbohydrate is degraded to provide energy and the first step in the metabolism is the cleavage of the molecule by phosphorylase. The enzyme catalyses the removal of one glucose unit at a time from the straight chain parts of the molecule and the formation of glucose-1-phosphate (G-1-P) from inorganic phosphate:

$$(\text{Glucose})_n + P_i \xrightarrow{\text{Phosphorylase}} (\text{Glucose})_{n-1} + \text{Glucose-1-phosphate}$$

Cleavage or lysis of the glycosidic link by water is known as *hydrolysis* and by phosphate as *phosphorolysis* hence the name of the enzyme *phosphorylase*. The reaction is readily reversible but *in vivo* the concentration of inorganic phosphate is such that the equilibrium favours degradation and not synthesis.

The reaction mixture contains cysteine to maintain the thiol groups (—SH) of the enzyme protein in a reduced form while fluoride is present to inhibit any phosphoglucomutase which would otherwise catalyse the conversion of G-1-P to G-6-P. The AMP is added for maximum enzyme activity as phosphorylase actually exists in two forms *a* and *b* and the *b* form is only active in the presence of AMP.

The G-1-P produced is isolated as the barium salt and, being a hemiacetal, is readily hydrolysed by dilute acid (0.5 mol/litre H_2SO_4) to glucose and phosphate. The reaction is virtually complete in 10 min at 100°C and this is used to check the stoichiometry of the product.

MATERIALS

		100
1.	Glycogen phosphorylase	25 units
2.	Sodium-β-glycerophosphate buffer (10 mmol/litre pH 6.8 containing 3 mmol/litre cysteine and 0.1 mol/litre NaF)	500 ml
3.	Glycogen (1.6 per cent w/v in 0.26 mol/litre phosphate buffer pH 6.8 containing 6 mmol/litre cysteine, 2 mol/litre AMP, and 0.2 mol/litre NaF)	500 ml
4.	Amylase	100 units
5.	Barium acetate (200 g/litre)	250 ml
6.	Phenol red indicator (10 g/litre in aqueous ethanol)	50 ml
7.	Sodium hydroxide (2 mol/litre)	500 ml
8.	Sulphuric acid (1 mol/litre)	500 ml
9.	Absolute ethanol	1 litre
10.	Reagents for phosphate estimation (Exp. 7.3)	—
11.	Boiling water bath	50
12.	Water bath at 37°C	20
13.	Measuring cylinder (25 ml)	50
14.	Crushed ice	—
15.	Reagents for the estimation of reducing sugars (Exp. 9.5)	—
16.	Glucose-1-phosphate (10 mmol/litre)	100 ml
17.	Glucose-6-phosphate (10 mmol/litre)	100 ml

METHOD

Incubation   Mix 5 ml of the β-glycerophosphate buffer with 5 ml of the glycogen solution in a boiling tube and equilibrate at 37°C for 10 min. Start the reaction by adding 0.2 ml of the phosphorylase, suitably diluted in the β-glycerophosphate buffer. Incubate for 30 min at 37°C and stop the reaction by placing the test tube into a boiling water bath for 2 min.

Isolation   Cool the tube under running tap water and remove the excess glycogen by adding 1 ml of the amylase solution and leaving at room temperature for 30 min. After the incubation with amylase, add 2 ml of the barium acetate solution and two drops of phenol red indicator followed by 1 mol/litre NaOH added dropwise until pink. Again mix thoroughly and remove the precipitate of barium phosphate by centrifuging at top speed on a bench centrifuge. Carefully remove the supernatant, measure the volume, and place in a flask; now add three times the volume of absolute ethanol, mix well and leave to stand for 30 min on ice. The precipitate formed under these conditions is the barium salt of glucose-1-phosphate.

Decant the supernatant and centrifuge the suspension for 10 min at top speed on a bench centrifuge. Remove the supernatant and allow the precipitate to drain by inverting the tubes on to filter paper. Repeat the precipitation procedure after adding 4 ml of water to the precipitated sugar phosphate.

Stoichiometry   Dissolve the barium salt in 5 ml of water and pipette 2 ml into a test tube containing 2 ml of 1 mol/litre H_2SO_4. Mix thoroughly and place in a boiling water bath for 10 min. Cool under running tap water, then add 2 ml of 2 mol/litre NaOH to neutralize the reaction; centrifuge and assay the supernatant for reducing sugar and phosphate using the methods in the book. Pipette another 2 ml sample of the phosphate ester into 2 ml of H_2SO_4, but this time add the NaOH immediately after and do not place the tube in a boiling water bath. This non-hydrolysed control is also assayed for phosphate and reducing sugar. Calculate the yield of G-1-P from the sugar and phosphate assays assuming complete hydrolysis.

Compare the results with those obtained after the acid hydrolysis of 1 mmol/litre solutions of pure glucose-1-phosphate and glucose-6-phosphate diluted from 10 mmol/litre stock solutions.

Further reading

Furth, A. J., *Lipids and Polysaccharides in Biology*. Arnold, London, 1980.
Ginsburg, V., *Biology of Carbohydrates*, Vol. 1. Wiley, New York, 1981.
Ginsburg, V. and Robbins, P. W., *Biology of Carbohydrates*, Vol. 2. Wiley, New York, 1984.
Guthrie, R. D., *Introduction to Carbohydrate Chemistry*. Clarendon Press, Oxford, 1974.

10. Lipids

Classification and biological role of lipids

Lipids are naturally occurring compounds that are esters of long chain fatty acids. They are insoluble in water but soluble in 'fat solvents' such as acetone, alcohol, chloroform or ether. Alkaline hydrolysis (known as saponification) gives rise to the alcohol and the sodium or potassium salts of the constituent fatty acids; these products of hydrolysis may be water-soluble. Chemically, lipids can be divided into two main groups: *simple lipids* and *compound lipids*. Steroids and the fat-soluble vitamins are also considered as lipids because of their similar solubility characteristics: they are known as *derived lipids*. However, many of these latter compounds are alcohols and not esters and hence cannot be saponified.

Simple lipids

Acylglycerols Esters of glycerol and fatty acids are known as *acylglycerols* or *glycerides*. The trihydric alcohol glycerol can be esterified to give mono-, di- and triglycerides. The fatty acids may be the same or different and, on saponification, free glycerol and fatty acids are obtained:

$$
\begin{array}{lll}
CH_2O \cdot COR_1 & CH_2OH & R_1COOK \\
| & | & \\
CH \cdot OCOR_2 + 3KOH \longrightarrow & CHOH + & R_2COOK \\
| & | & \\
CH_2O \cdot COR_3 & CH_2OH & R_3COOK \\
\text{Triglyceride} & \text{Glycerol} & \text{K salts of} \\
& & \text{fatty acids}
\end{array}
$$

Triglycerides are the predominant form in nature, although mono- and diglycerides are known. The acylglycerols are uncharged molecules and for this reason are also known as *neutral lipids*. They are called *fats* or *oils* depending on whether they are solid or liquid at room temperature.

If the fatty acids substituted at positions 1 and 3 are different then C-2 becomes a chiral centre and two stereoisomers are possible, although most triglycerides in nature are of the L form.

Fatty acids A wide range of fatty acids are found in nature and some of the commonest ones are given below (Table 10.1). Most of the natural fatty acids have straight chains with an even number of carbon atoms, although branch chain and

Table 10.1   Some common fatty acids found in living organisms

Trivial name	Formula	Symbol
Saturated fatty acids		
Lauric	$CH_3(CH_2)_{10}COOH$	12:0
Myristic	$CH_3(CH_2)_{12}COOH$	14:0
Palmitic	$CH_3(CH_2)_{14}COOH$	16:0
Stearic	$CH_3(CH_2)_{16}COOH$	18:0
Unsaturated fatty acids		
Oleic	$CH_3(CH_2)_7CH{=}CH(CH_2)_7COOH$	$18{:}1^{\Delta,9}$
Linoleic	$CH_3(CH_2)_4CH{=}CH \cdot CH_2 \cdot CH{=}CH(CH_2)_7COOH$	$18{:}2^{\Delta,9,12}$
Linolenic	$CH_3CH_2CH{=}CH \cdot CH_2 \cdot CH{=}CH \cdot CH_2 \cdot CH{=}CH(CH_2)_7COOH$	$18{:}3^{\Delta,9,12,15}$
Arachidonic	$CH_3(CH_2)_4(CH{=}CH\ CH_2)_3CH{=}CH(CH_2)_3COOH$	$20{:}4^{\Delta,5,8,11,14}$

cyclic fatty acids are not unknown. Many of the fatty acids are unsaturated molecules and the introduction of a double bond into the fatty acid part of an acylglycerol lowers the melting point of the compound; thus, animal fats, which consist largely of triglycerides with fully saturated fatty acids, are solid at room temperature, while vegetable and fish oils, which contain a high proportion of unsaturated fatty acids, are liquid at room temperature. Furthermore, the introduction of a double bond gives rise to two possible geometric forms, the *cis* and *trans* isomers, although the *trans* form is rare in nature.

$$\Delta^9\text{-Octadecanoic acid}$$

$$\begin{array}{ll}
CH(CH_2)_7CH_3 & CH_3(CH_2)_7\,CH \\
\parallel & \parallel \\
CH(CH_2)_7\,COOH & CH(CH_2)_7\,COOH \\
\textit{cis} & \textit{trans} \\
\text{Oleic acid} & \text{Elaidic acid}
\end{array}$$

A simple notation for fatty acids is given in Table 10.1. The first number denotes how many carbon atoms are in the chain, and the second one the number of double bonds. The superscript numbers following Δ indicate the position of any double bonds in the molecule. Stearic acid, for example, is shown as 18:0 which indicates 18 carbon atoms and no double bonds. Linoleic acid is given as $18{:}2^{\Delta,9,12}$ which shows 18 carbon atoms and two double bonds at positions 9–10 and 12–13.

Triacylglycerols (*triglycerides*)   These compounds make up the bulk of ingested lipid. They are partially degraded by lipases in the gut then re-esterified in the gut mucosa. The fat is then transported to the blood via the lymphatic system in the form of chylomicrons; these are fat droplets from 0.1 to 1 μm in diameter made up largely of triglycerides with some cholesterol and a lipoprotein skin. This lipid may then be oxidized in the liver to provide energy or deposited as depot fat, in characteristic regions of the animal, where it acts as a long-term food store and heat insulator. In some seeds, the triglycerides are stored in the form of an oil.

Compound lipids

Complete hydrolysis of a compound lipid yields at least one other component as well as the usual alcohol and fatty acids. These compounds are essential structural components of cell membranes and this is discussed in detail later.

Phosphoglycerides (*glycerol phosphatides*) These compounds are also known as *phospholipids* and are very abundant in living organisms. They are chemically similar to the triacylglycerols, being fatty acid esters of glycerol, but in addition they contain phosphoric acid esterified with an alcohol (X). The general formula of these compounds is:

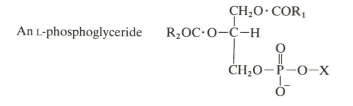

An L-phosphoglyceride

Carbon-2 of the glycerol is a chiral centre and the naturally occurring compounds are part of the L series as given in the above example.

Table 10.2 Some naturally occurring phospholipids

Recommended name	Alcohol moiety X	Trivial name
Choline phosphoglyceride	$-O-CH_2CH_2\overset{+}{N}(CH_3)_3$	Lecithin
Ethanolamine phosphoglyceride	$-O-CH_2CH_2NH_2$	Cephalin
Serine phosphoglyceride	$-O-CH_2CH(NH_2)COOH$	—
Inositol phosphoglyceride		—

Four of the commonest phospholipids are shown in Table 10.2. Their names are derived from the alcohol X linked to the phosphoric acid residue. (The recommended name is the alcohol followed by phosphoglyceride or phosphatidyl linked to the name of the alcohol.) One of the most abundant phospholipids contains choline so this is known as *choline phosphoglyceride* or *phosphatidylcholine*.

Plasmalogens The plasmalogens have a very similar structure to the phosphoglycerides except the fatty acid in the 1 position is joined to the glycerol through a vinyl ether bond and not the usual ester link.

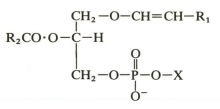

A plasmalogen

Sphingolipids   The sphingolipids contain the amino alcohol sphingosine or a related compound as a backbone instead of glycerol.

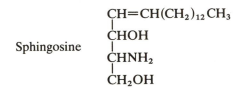

Sphingosine

In the case of the *sphingomyelins*, the primary alcohol group of sphingosine is esterified with phosphatidyl ethanolamine and the fatty acid residue is present as the acyl derivative of the amino group. Sphingolipids are generally metabolized more slowly than the phospholipids and appear to make up the more stable structures of cells. For example, the sphingomyelin of the myelin sheath which acts as an insulator for nerve fibres is not metabolized in adult life.

Glycolipids   As the name suggests, these compounds contain both carbohydrate and lipid moieties. *Cerebrosides* have a similar structure to sphingomyelin except that the sugar galactose replaces the phosphoryl choline. The *gangliosides* are again derivatives of sphingosine but in this case an oligosaccharide chain containing *N*-acetylneuraminic acid is joined to the primary alcohol group of sphingosine. Cerebrosides are present in high concentrations in nervous tissue, particularly in the white matter of brain, while the gangliosides on the other hand are found in high concentrations in the grey matter.

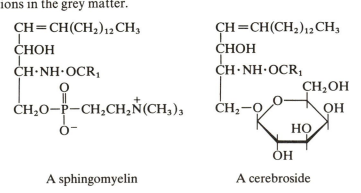

A sphingomyelin   A cerebroside

Lipoproteins   Lipids in biological material are frequently associated with proteins as lipoprotein complexes. The soluble lipoproteins are concerned with the transport of lipids in the blood, while insoluble lipoproteins constitute the main part of many biological membranes such as the endoplasmic reticulum, the lamellae of chloroplasts, the cristae of mitochondria and the myelin sheath of nerves.

Derived lipids

Steroids   These compounds are soluble in the usual lipid solvents, but most of them are not saponified. They are usually considered along with lipids because of their similar solubility characteristics. All steroids have the 17 carbon perhydrocyclopentanophenanthrene ring as the basis of their structure.

Perhydrocyclopentanophenanthrene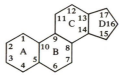

When a substituent lies *above* the plane of the ring, it is known as β and is indicated by a full line, whereas if it is situated *below* the ring the letter α is used and the position of the group is denoted by a dotted line.

Steroids are of widespread occurrence in higher animals where they perform a variety of functions. For example, *cholesterol*, which is the most abundant of the steroids, is present in many animal cell membranes and serves as an important precursor of many other steroids.

Cholesterol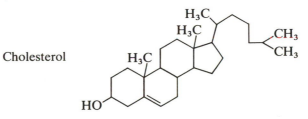

The bile acids *cholic acid* and *deoxycholic*, involved in the digestion and adsorption of fats, are steroids and so is the hormone *aldosterone*, from the adrenal cortex, which regulates water and electrolyte balance in mammals. Many other hormones are steroids and only slight changes in their chemical structure profoundly affect their biological activity, as in the case of the sex hormones shown below.

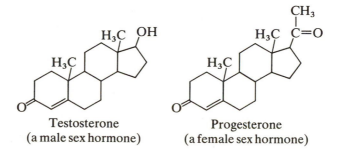

Testosterone
(a male sex hormone)

Progesterone
(a female sex hormone)

Fat-soluble vitamins   These compounds are found in association with natural lipid foods and are included with lipids because of their solubility properties (Table 10.3).

Table 10.3   Some fat-soluble vitamins

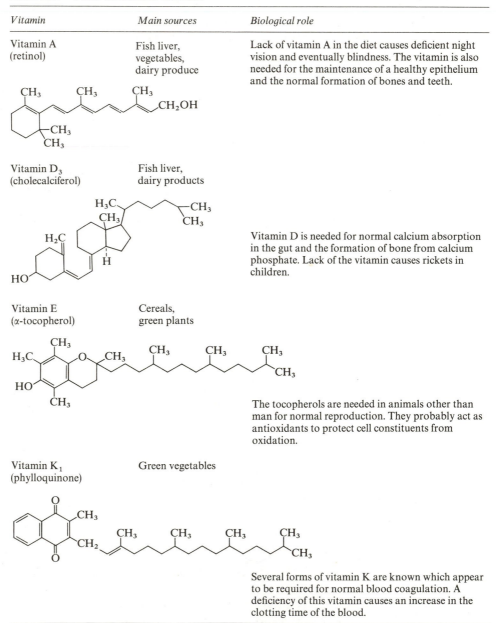

Vitamin	Main sources	Biological role
Vitamin A (retinol)	Fish liver, vegetables, dairy produce	Lack of vitamin A in the diet causes deficient night vision and eventually blindness. The vitamin is also needed for the maintenance of a healthy epithelium and the normal formation of bones and teeth.
Vitamin D_3 (cholecalciferol)	Fish liver, dairy products	Vitamin D is needed for normal calcium absorption in the gut and the formation of bone from calcium phosphate. Lack of the vitamin causes rickets in children.
Vitamin E (α-tocopherol)	Cereals, green plants	The tocopherols are needed in animals other than man for normal reproduction. They probably act as antioxidants to protect cell constituents from oxidation.
Vitamin K_1 (phylloquinone)	Green vegetables	Several forms of vitamin K are known which appear to be required for normal blood coagulation. A deficiency of this vitamin causes an increase in the clotting time of the blood.

Quantitative analysis of lipids

A complete chemical analysis of a naturally occurring fat is quite a lengthy procedure, but there are a number of measurements such as the acid value, the saponification number and the iodine number, which give useful information on the composition and purity of a particular fat.

Experiment 10.1   The determination of the acid value of a fat

PRINCIPLE

During storage, fats may become rancid as a result of peroxide formation at the double bonds by atmospheric oxygen and hydrolysis by micro-organisms with the liberation of free acid. The amount of free acid present therefore gives an indication of the age and quality of the fat.

It is suggested that each pair of students choose one lipid and assay (a) a fresh sample and (b) one that has been exposed to the atmosphere. Results should then be compared with other groups who have determined other lipids.

The acid value is the number of milligrams of KOH required to neutralize the free fatty acid present in 1 g of fat.

MATERIALS	$\underline{100}$
1. Olive oil, butter and margarine (use a fresh sample and one that has stood for several days at room temperature)	200 g
2. Fat solvent (equal volumes of 95 per cent v/v alcohol and ether neutralized to phenolphthalein)	6 litres
3. Phenolphthalein (10 g/litre in alcohol)	200 ml
4. Potassium hydroxide (0.1 mol/litre)	1 litre
5. Burettes (5 ml and 25 ml)	50

METHOD

Accurately weigh out 10 g of the test compound and suspend the melted fat in about 50 ml of fat solvent. Add 1 ml of phenolphthalein solution, mix thoroughly, and titrate with 0.1 mol/litre KOH until the faint pink colour persists for 20–30 s. Note the number of millilitres of standard alkali required and calculate the acid value of the fat. *Note:* 0.1 mol/litre KOH contains 5.6 g/litre or 5.6 mg/ml.

Experiment 10.2   The saponification value of a fat

PRINCIPLE

On refluxing with alkali, glyceryl esters are hydrolysed to give glycerol and the potassium salts of the fatty acids (soaps).

The materials are calculated on the assumption that each pair of students assay one fat or oil.

The saponification value is the number of milligrams of KOH required to neutralize the fatty acids resulting from the complete hydrolysis of 1 g of fat. The saponification value gives an indication of the nature of the fatty acids in the fat since the longer the carbon chain the less acid is liberated per gram of fat hydrolysed.

MATERIALS   100
1. Fats and oils (tristearin, coconut oil, corn oil, and butter)   20 g
2. Fat solvent (equal volumes of 95 per cent ethanol and ether)   1 litre
3. Alcoholic KOH (0.5 mol/litre)   3 litres
4. Reflux condenser   100
5. Boiling water bath   100
6. Phenolphthalein (10 g/litre in alcohol)   50 ml
7. Hydrochloric acid (0.5 mol/litre)   3 litres
8. Burettes (10 ml and 25 ml)   100
9. Conical flasks (250 ml)   100

METHOD
Accurately weigh 1 g of the fat in a tared beaker and dissolve in about 3 ml of the fat solvent. Quantitatively transfer the contents of the beaker to a 250 ml conical flask by rinsing the beaker three times with a further millilitre of solvent; add 25 ml of 0.5 mol/litre alcoholic KOH and attach to a reflux condenser. Set up another reflux condenser as blank with everything present except the fat and heat both flasks on a boiling water bath for 30 min. Leave to cool to room temperature and titrate with 0.5 mol/litre HCl and phenolphthalein indicator.

The difference between the blank and test reading gives the number of millilitres of 0.5 mol/litre KOH required to saponify 1 g of fat.

The molecular weight of KOH is 56 and, since three molecules of fatty acid are released from a triglyceride, then:

$$\text{Saponification value } (S) = 3 \times 56 \times 1000/\text{Average mol. wt of fat}$$

$$\therefore \text{ Average mol. wt of fat} = 3 \times 56 \times 1000/S$$

Experiment 10.3   The iodine number of a fat

PRINCIPLE
Halogens add across the double bonds of unsaturated fatty acids to form addition compounds.

Iodine monochloride (ICl) is allowed to react with the fat in the dark. The amount of iodine consumed is then determined by titrating the iodine released (after adding KI) with standard thiosulphate and comparing with a blank in which the fat is omitted.

$$-CH=CH- \; + \; ICl \longrightarrow \quad \begin{matrix} H & H \\ | & | \\ -C-C- \\ | & | \\ I & Cl \end{matrix}$$

$$ICl \; + \; KI \longrightarrow HCl \; + \; I_2$$

$$I_2 \; + \; 2Na_2S_2O_3 \longrightarrow 2NaI \; + \; Na_2S_4O_6$$

The reaction mixture is kept in the dark and the titration carried out as quickly as possible since halogens are oxidized in the light. The quantity of material is calculated assuming that each pair of students will assay two of the fats.

The iodine number is the number of grams of iodine taken up by 100 g of fat.

MATERIALS
		100
1.	Fats (20 g/litre solutions of corn oil, olive oil, linseed oil, and butter in chloroform)	1 litre
2.	Iodine monochloride (0.2 mol/litre approx.)	3 litres
3.	Potassium iodide (100 g/litre)	1.5 litres
4.	Sodium thiosulphate (0.1 mol/litre)	4 litres
5.	Starch indicator (10 g/litre)	250 ml
6.	Stoppered bottles (250 ml)	200
7.	Burette (25 ml)	100
8.	Chloroform	1 litre

METHOD

Pipette 10 ml of the fat solution into a stoppered bottle, add 25 ml of the ICl solution, stopper the bottle, and leave to stand in the dark for 1 h, after shaking thoroughly. At the same time, set up a blank in which the fat solution is replaced by 10 ml of chloroform.

Rinse the stoppers and necks of the bottles with about 50 ml of water, add 10 ml of the KI solution, and titrate the liberated iodine with the standard thiosulphate. When the solution is a pale straw colour, add about 1 ml of starch solution and continue titrating until the blue colour disappears. The bottles must be shaken thoroughly throughout the titration to ensure that all the iodine is removed from the chloroform layer.

The difference between the blank and test readings $(Bl - T)$ gives the number of ml of 0.1 mol/litre thiosulphate needed to react with the equivalent volume of iodine. This is $(Bl - T)/2$ ml of 0.1 mol/litre iodine since 2 molecules of thiosulphate are needed for each iodine. The mol. wt of iodine is 2×127 so the weight of iodine in $(Bl - T)/2$ ml of 0.1 mol/litre iodine is:

$$(Bl - T)/2 \times 0.1 \times 2 \times 127/1000 g$$

The amount of fat taken was 0.2 g so the iodine number is:

$$(Bl - T) \times 12.7/1000 \times 100/0.2$$

Iodine number $= (Bl - T) \times 6.35$ g per 100 g of fat.

Experiment 10.4   The estimation of blood cholesterol

PRINCIPLE

Acetic anhydride reacts with cholesterol in a chloroform solution to produce a characteristic blue–green colour. The exact nature of the chromophore is not known but the reaction probably includes esterification of the hydroxyl group in the 3 position as well as other rearrangements in the molecule.

Blood or serum is extracted with an alcohol–acetone mixture which removes cholesterol and other lipids and precipitates protein. The organic solvent is removed by evaporation on a boiling water bath and the dry residue dissolved in chloroform. The cholesterol is then determined colorimetrically using the Liebermann–Burchard reaction.

Free cholesterol is equally distributed between the cells and plasma while the esterified form occurs only in plasma.

It is essential to use absolutely *dry* glassware for this estimation.

MATERIALS	100
1. Serum or blood	25 ml
2. Alcohol–acetone mixture (1:1)	2 litres
3. Chloroform	500 ml
4. Acetic anhydride–sulphuric acid mixture (30:1 mix just before use, *Care*!)	930 ml
5. Stock cholesterol solution (2 mg/ml in chloroform)	250 ml
6. Working cholesterol solution. (Dilute the above solution one in five with chloroform to give a solution of 0.4 mg/ml)	1 litre

METHOD

Place 10 ml of the alcohol–acetone solvent in a centrifuge tube and add 0.2 ml of serum or blood. Immerse the tube in a boiling water bath with shaking until the solvent begins to boil. Remove the tube and continue shaking the mixture for a further 5 min. Cool to room temperature and centrifuge. Decant the supernatant fluid into a test tube and evaporate to dryness on a boiling water bath. Cool and dissolve the residue in 2 ml of chloroform. At the same time, set up a series of standard tubes containing cholesterol and a blank with 2 ml of chloroform.

Add 2 ml of acetic anhydride–sulphuric acid mixture to all tubes and thoroughly mix. Leave the tubes in the dark at room temperature and read the extinction at 680 nm.

APPLICATION

The normal serum cholesterol lies within the range 100–250 mg/100 ml. The average serum cholesterol is about 200 mg/100 ml at 25 in men and rises slowly with age, reaching a peak figure at age 40–50 and then declining. Women in general show a lower cholesterol level than men until the menopause, when the value rises above that found in men of the same age.

A high serum cholesterol of 300 mg/100 ml or more in young adults is a serious indication of coronary disease.

High levels of as much as 25 per cent above normal are also found in nephritis, diabetes, myxoedema, and xanthomatosis.

The separation and isolation of lipids

Experiment 10.5   The lipid composition of wheat grain

PRINCIPLE

The predominant food reserve in the wheat grain is starch. Lipid accounts for only 2–4 per cent of the dry weight of the whole grain and most of this is concentrated in the germ or embryo. The major lipid component is triglyceride and this is hydrolysed during germination by the enzyme lipase. The products of enzymic hydrolysis are then utilized by the growing shoot.

The following experiment demonstrates the extraction, separation and composition of the wheat lipids using thin layer chromatography to identify the main components. The lipid compositions of whole wheat grain, wheat germ and a vegetable oil (olive, sunflower, etc.) are compared.

MATERIALS	10
1. Wheat grain	—
2. Commercial wheat germ	—
3. Chloroform-methanol extraction mixture (2:1)	90 ml
4. Water bath at 50°C	5
5. Nitrogen cylinders	2
6. Olive oil and sunflower oil	10 ml
7. Materials for TLC of lipids (Exp. 4.6)	—
8. Pestle and mortar	5

METHOD

Extraction   Crush five wheat grains and extract the lipids by grinding in a pestle and mortar for several minutes with 2 ml of the chloroform–methanol mixture. Remove the debris by centrifugation and transfer the supernatant to a test tube in a water bath at 50°C and evaporate to a small volume (0.2 ml) under a stream of nitrogen in a fume chamber. Extract the lipids in a similar way from a commercial sample of wheat germ.

Thin layer chromatography   Spot samples of whole wheat lipids, wheat germ lipids and olive or sunflower oil on to thin layer plates together with the standards. Run the plates in the chromatography solvent and identify the lipids as far as possible using the system described in Exp. 4.6.

Further experiments   Examine the lipid composition of wheat grain after germination for a week.

Compare the lipid composition of wheat seedlings grown in the dark with those grown in the light.

Experiment 10.6   The preparation of cholesterol from brain

PRINCIPLE

Cholesterol is readily soluble in acetone, while most complex lipids are insoluble in this solvent.

MATERIALS	10
1. Brains from calf or pig	1 kg
2. Acetone	3.5 litres
3. Waring blender	3
4. Large Büchner filtration apparatus	5
5. Distillation equipment	5
6. Large filter funnel and fluted paper	5
7. Boiling water bath	5
8. Ethanol	1 litre

METHOD

Add 400 ml of acetone to 100 g of brain and blend for 1 min. Rinse out the blender with a little acetone and stir the combined homogenates for 10 min. Filter the suspension on a Büchner funnel, blend the residue with a further 200 ml of acetone as before, filter and combine the filtrates. After removing most of the acetone by distillation under reduced pressure, cool the flask with tap water and collect the crude cholesterol on a Büchner funnel.

Dissolve the crude material in the minimum volume of hot ethanol and filter while hot using a fluted filter paper. For this, the collecting flask is placed in a boiling water bath with the filter funnel in position.

Air dry the cholesterol and record the yield and melting point. If necessary, the cholesterol may be crystallized from hot ethanol as above.

Determine the purity of your product by preparing a standard solution in chloroform and estimating the amount of cholesterol present by the Liebermann–Burchard reaction (Exp. 10.4).

Fat-soluble vitamins

Vitamins are necessary food factors required for the maintenance of health and their absence from the diet leads to a number of deficiency diseases (Table 10.3). In some ways vitamins are similar to hormones in that only small quantities may be needed to produce a large physiological effect. As with hormones, the vitamins are chemically diverse and perform a wide range of functions. Two of the fat-soluble vitamins (A and D) and their interaction with ultraviolet radiation are dealt with in the next two experiments.

Experiment 10.7   The effect of ultraviolet light on vitamin A

PRINCIPLE

Vitamin A absorbs in the ultraviolet region of the spectrum with a maximum at 325 nm, and this property can be used for the estimation of the vitamin. The method is rapid and sensitive but suffers from the disadvantage of low specificity. A number of other substances absorb in the same region of the spectrum as vitamin A and allowance has to be made for these interfering compounds. One method is to apply a correction formula to check for the presence of interfering compounds and to correct for their absorption in the region of 325 nm. For example, pure vitamin A alcohol in isopropanol gives six-sevenths of the maximum extinction at 310 nm and 334 nm so that, if the extinction at 325 nm is scaled to 1.00, then the absorbance at 310 nm and 334 nm becomes 0.857. The corrected extinction at 325 nm (E_{corr}) can be obtained by the application of the formula:

$$E_{corr} = 7E_{325} - 2.625E_{310} - 4.375E_{334}$$

Another way of detecting and allowing for these compounds is to measure the extinction at 325 nm before and after irradiating with ultraviolet light. Vitamin A is destroyed by ultraviolet light and it is assumed that the interfering chromagens are unaffected by this treatment. The fall in extinction following irradiation is therefore a direct measure of the vitamin A content of the sample.

MATERIALS   10
1 Isopropanol   50 ml
2. Vitamin A standard solution (1 µg/ml in isopropanol, keep in a sealed   5 ml
   bottle under nitrogen in the dark)
3. Cod liver oil and halibut liver oil   50 ml
4. Ultraviolet spectrophotometer   3
5. Ultraviolet lamp   5

METHOD

Samples such as fish oils which contain high concentrations of the vitamin can be diluted directly with isopropanol, although probably more accurate values are obtained when the material is saponified and the vitamin A, which is unsaponifiable , is extracted.

Measure the extinction of a freshly diluted sample of the fish oil in isopropanol at 325 nm against a solvent blank (T_1), remove the cuvette, expose to ultraviolet light until the extinction no longer falls with time, and again record the absorbance (T_2). Treat the standard vitamin A solution in the same way (St_1 and St_2) and calculate the vitamin A concentration of the sample:

$$\text{Vitamin A concentration (µg/ml)} = \frac{T_1 - T_2}{St_1 - St_2} \times 1 \times \text{Dilution factor.}$$

Record the absorption spectrum of the freshly diluted test sample from 280 to 380 nm and compare this with the absorption spectrum of the standard vitamin A solution by plotting E/E_{325} for the standard and test samples.

What is the purity of the vitamin A standard assuming the correction formula to be accurate?

Do fish oils contain many compounds that interfere with the assay procedure?

How do the results of this method compare with those obtained by irradiating the test samples?

Experiment 10.8   Preparation of the D vitamins by irradiation of their precursors with ultraviolet light

PRINCIPLE

Ergosterol, which occupies a similar place in plants to that of cholesterol in animals, is an important precursor of vitamin D_2. Irradiation of ergosterol (provitamin D_2) with ultraviolet light causes a number of rearrangements to take place in the molecule forming provitamin D_2 and finally calciferol (vitamin D_2). Lumisterol and tachysterol are also formed as by-products of this photochemical reaction (Fig. 10.1).

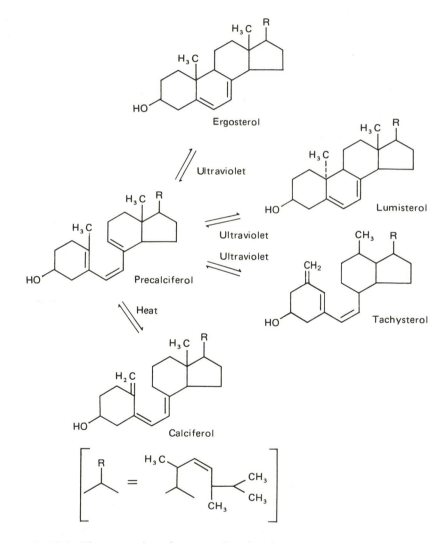

Fig. 10.1   The conversion of ergosterol to vitamin D_2

Similar rearrangements take place when 7-dehydrocholesterol is irradiated giving provitamin D_3 and cholecalciferol (vitamin D_3). This reaction takes place in the skin of man and other mammals exposed to sunlight or ultraviolet radiation.

Gentle warming of a solution of calciferol in the absence of light and in an atmosphere of nitrogen produces an equilibrium mixture of provitamin D_3 and calciferol.

MATERIALS

		<u>10</u>
1.	Thin layer plates of silica-gel G	20
2.	Separation chambers	5
3.	Ergosterol and dehydrocholesterol	1 g
4.	Chromatography solvent (chloroform)	2 litres
5.	Chromatography standards related to vitamin D_2 (ergosterol, provitamin D_2, tachysterol, and lumisterol in chloroform, approx. 20 mg/ml)	2 ml
6.	Chromatography standards related to vitamin D_3 (dehydrocholesterol, provitamin D_3, vitamin D_3, and cholesterol in chloroform, approx. 20 mg/ml)	2 ml
7.	Nitrogen cylinder	2
8.	Fluorimeter	5
9.	Developing reagent (5 per cent v/v sulphuric acid)	1 litre
10.	Oven at 140°C	3
11.	Apparatus for spraying thin layer chromatograms	5
12.	Silica cells	5
13.	Ultraviolet lamp	3

METHOD

Ultraviolet irradiation   Prepare a solution of ergosterol in chloroform (20–30 mg/ml), transfer an aliquot to a silica cuvette of a fluorimeter, flush out with nitrogen and seal. Place the cuvette in the fluorimeter and irradiate with ultraviolet light of short wavelength for 20 min. Remove 10 μl of the test solution and spot this on to a thin layer plate of silica-gel G. Rapidly spot out the standards related to vitamin D_2 on the thin layer plate and place immediately in the chloroform solvent.

Repeat the above test with dehydrocholesterol instead of ergosterol and identify the products formed.

Heat   Examine the effect of heating solutions of vitamin D_2 and vitamin D_3 in chloroform at 60°C in the absence of light and air for periods of time up to 4 h.

Detection of vitamins   Examine the plates under ultraviolet light when some of the compounds show up as fluorescent spots. Alternatively, the plates can be sprayed with dilute sulphuric acid (*Care!*), heated at 140°C for 5–10 min, then viewed in visible or ultraviolet light. Compare the sensitivity of the two methods of detection.

Further reading

Furth, A. J., *Lipids and Polysaccharides in Biology*. Arnold, London, 1980.

Gurr, M. I. and James, A. T., *Lipid Biochemistry: An Introduction*, 3rd edn. Chapman and Hall, London, 1980.

Mead, J. F., Alfin-Slater, R. B., Howton, D. R. and Popjak, G., *Lipids: Chemistry, Biochemistry and Nutrition*. Plenum Press, New York, 1986.

11. Nucleic acids

Chemical composition of the nucleic acids

Nucleic acids are nitrogen-containing compounds of high molecular weight found in association with proteins in the cell. The nucleic acid–protein complexes are known as nucleoproteins, and these can be separated into the component proteins and nucleic acids by treatment with acid or high salt concentration. The proteins are basic in character and the nucleic acids, as the name suggests, are acidic. Two main groups of nucleic acids are known, ribonucleic acid (RNA) and deoxyribonucleic acid (DNA). Hydrolysis of DNA and RNA under controlled conditions yields *nucleotides*, which can be regarded as the basic unit of nucleic acids, just as amino acids are the basic unit of proteins and monosaccharides of polysaccharides. Further hydrolysis of the nucleotides yields *nucleosides* and eventually phosphate, a sugar, and a number of purine and pyrimidine bases. The relationship between these components and the nucleoprotein is shown for DNA (Fig. 11.1).

RNA has a similar composition, except that the sugar ribose is present instead of deoxyribose and uracil instead of thymine.

The formulae of the main nitrogen bases found in DNA and RNA are shown below.

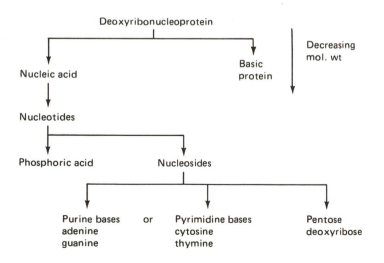

Fig. 11.1 The relationship between deoxyribonucleoprotein and its low molecular weight components

Purines

Adenine and guanine are substituted purines and are present in all nucleic acids.

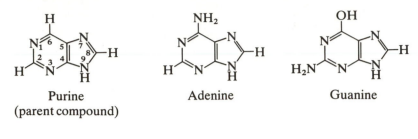

Purine
(parent compound)   Adenine   Guanine

Pyrimidines

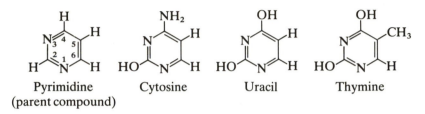

Pyrimidine   Cytosine   Uracil   Thymine
(parent compound)

Small quantities of other bases have been detected in nucleic acids from some sources. It should be noted that both purine and pyrimidine bases can exist in the keto or enol form (Fig. 11.2).

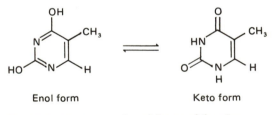

Enol form   Keto form

Fig. 11.2   The keto and enol forms of thymine

The pentose sugars

The two main groups of nucleic acids derive their names from the sugar present, which is either ribose or deoxyribose.

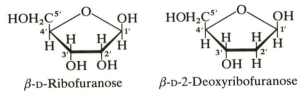

β-D-Ribofuranose   β-D-2-Deoxyribofuranose

The carbon atoms on the sugars are denoted as 1′, 2′, etc., in order to differentiate them from the atoms of the bases.

Nucleosides

The C-1 of the sugar is linked to the nitrogen at the 9 position for purines or 1 position for pyrimidines to form a nucleoside.

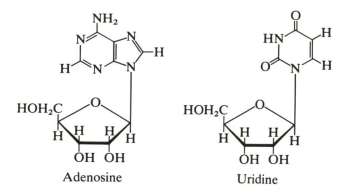

Adenosine   Uridine

Most of the bonds linking the sugar and the base are as shown above, but transfer RNA does contain an unusual nucleotide, pseudouridine, in which ribose is linked to the uracil via the 5 position.

Nucleotides

The hydroxyl groups on positions 2', 3' and 5' of ribose can be esterified with phosphoric acid and all these esters are known. Similarly, the 3' and 5' positions of deoxyribose can be esterified and these esters are also known to exist.

Nucleotides and nucleosides are named after the bases contained in their structure, as below.

Base	Nucleoside	Nucleotide
Adenine	Adenosine	Adenylic acid
Guanine	Guanosine	Guanylic acid
Uracil	Uridine	Uridylic acid
Cytosine	Cytidine	Cytidylic acid
Thymine	Thymidine	Thymidylic acid

If the base is linked to deoxyribose, then the names are modified so that a nucleoside consisting of adenine and deoxyribose would be called deoxyadenosine. As well as the nucleotides indicated, a number of biologically important nucleotides such as adenosine di- and triphosphate (ADP and ATP), guanosine di- and triphosphate (GDP and GTP), and nicotinamide adenine dinucleotide (NAD) occur in the free state.

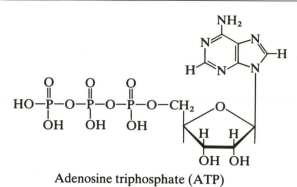

Adenosine triphosphate (ATP)

Nucleic acids

Nucleic acids are macromolecules in which the nucleotides are linked by phosphodiester bonds between the 3′ and 5′ positions of the sugars.

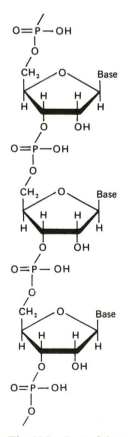

Fig. 11.3   Part of the molecule of RNA

A portion of a molecule of RNA therefore has the structure given in Fig. 11.3 where the base is either a purine or pyrimidine.

Most nucleic acids are very large molecules, so that to show the complete formulae would be rather cumbersome. A useful form of shorthand for the structure shows the bases present by using their first letter. The sequence of bases is extremely important so that a nucleic acid may be shown by the first letters of the bases only (Fig. 11.4).

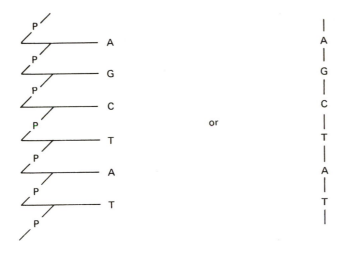

Fig. 11.4   Shorthand notation to show the base sequence of a nucleic acid

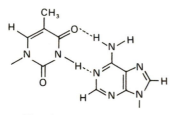

Thymine   Adenine

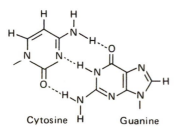

Cytosine H   Guanine

Fig. 11.5   The pairing of bases by hydrogen bonding (—) as in DNA

DNA   DNA consists of two chains of polynucleotides interwoven in the form of a spiral structure which is stabilized by hydrogen bonding between particular base pairs. The stereochemistry of the bases is such that adenine pairs with thymine and guanine with cytosine so that the ratio of A/T and G/C is unity (Fig. 11.5). This aspect of the structure of DNA was first proposed by Watson and Crick in 1953 on the basis of the X-ray data of Wilkins and is known as the Watson–Crick hypothesis. This idea has since been confirmed and provides the basis for explaining some of the biological properties of DNA.

Most DNA molecules are of the double helical type, although some viruses contain only single-stranded DNA and mitochondrial DNA is a closed loop.

It is difficult to arrive at an accurate value for the molecular weight of DNA since the methods used for its isolation may result in breaks in the molecule, but values of 10^9 have been obtained.

RNA   The simple relationship of A/U = G/C = 1 does not hold for most forms of RNA, since the RNA molecule consists of a single strand of nucleic acid in the form of a random coil with only limited regions of base pairing. The molecular weight of transfer RNA is about 25 000, but other forms of RNA have a high molecular weight of a million or so.

The biological role of nucleic acids

This branch of biochemistry has grown rapidly during recent years and it is not possible to do it justice in only a few words, but a brief outline of the place of these molecules in living organisms would seem appropriate at this point.

DNA

Occurrence   DNA is intimately associated with the genetic material of the cell. In some micro-organisms, a single strand of DNA seems to be the store of the genetic information but, in higher organisms, the DNA is present as nucleoprotein in the chromosomes. The amount of DNA in the cell of a particular species is constant, whereas the germ cells, with half the number of chromosomes, contain half the amount of DNA present in other cells. The DNA occurs almost exclusively in the nucleus with trace amounts in the mitochondria and chloroplasts.

Function   The main function of DNA is to act as a store of genetic information. Hereditary characteristics are passed on to daughter cells through replication of DNA which appears to take place in a series of short steps as segments of the double helix are unwound and the new DNA is formed on each strand simultaneously from free nucleotides. The newly synthesized strands are built up according to the rules on base pairing so that eventually two identical molecules are formed from the original DNA.

DNA is also the template for the synthesis of proteins in the cell with a triplet of three bases providing the *genetic code* for each amino acid.

RNA

Occurrence   Whereas DNA is found almost exclusively in the nucleus, RNA is distributed throughout the cell. Most of the RNA is present in the cytoplasm as soluble and ribosomal RNA, but about 10 per cent is found in the nucleus with trace amounts also present in the mitochondria. There are three types of RNA present in the cells of higher organisms: *messenger RNA (mRNA)*, *ribosomal RNA (rRNA)* and *transfer RNA (tRNA)*, all of which are actively involved in the synthesis of proteins.

Protein synthesis   The genetic message is first passed from the nuclear DNA to mRNA through the bases, a process known as *transcription*. The mRNA then migrates into the cytoplasm and it is here in association with the ribosomes and tRNA that the four-letter base code is translated into the 20-letter amino acid code of proteins: a process known as *translation*. Each amino acid has its own tRNA whose function is to transfer the activited amino acid to the site of protein synthesis.

Nucleic acids of viruses

All viruses contain nucleic acids which may account for as much as 50 per cent of the particle. The nucleic acid may be single or double stranded and may be present as a linear or a circular molecule. The nucleic acid has a well-defined tertiary structure and is surrounded by a protective coat of protein. The nucleic acid is the infective part of the virus while the protein accounts for its immunological specificity. The virus attaches itself to the cell and injects its own DNA into the host, thus directing the infected cell to synthesize virus proteins and nucleic acids in order to multiply. Plant viruses contain RNA but no DNA, and in these cases the RNA acts as the genetic material as well as performing its usual function in protein synthesis.

Experiments

Experiment 11.1   The isolation of RNA from yeast

PRINCIPLE

Total yeast RNA is obtained by extracting a whole cell homogenate with phenol. The concentrated solution of phenol disrupts hydrogen bonding in the macromolecules, causing denaturation of the protein. The turbid suspension is centrifuged and two phases appear: the lower phenol phase contains DNA, and the upper aqueous phase contains carbohydrate and RNA. Denatured protein, which is present in both phases, is removed by centrifugation. The RNA is then precipitated with alcohol. The product obtained is free of DNA but usually contaminated with polysaccharide. Further purification can be made by treating the preparation with amylase.

MATERIALS	$\underline{10}$
1. Dried yeast	200 g
2. Phenol solution (900 g/litre)	1 litre

3. Potassium acetate (200 g/litre, pH 5)	200 ml
4. Absolute ethanol	3 litres
5. Diethyl ether	500 ml
6. Water bath at 37°C	5

METHOD

Suspend 30 g of dried yeast in 120 ml of water previously heated to 37°C. Leave for 15 min at this temperature and add 160 ml of concentrated phenol solution (*Care: corrosive*). Stir the suspension mechanically for 30 min at room temperature, then centrifuge at 3000 g for 15 min in the cold to break the emulsion. Carefully remove the upper aqueous layer with a Pasteur pipette and centrifuge at 10 000 g for 5 min in a refrigerated centrifuge to sediment denatured protein. Add potassium acetate to the supernatant to a final concentration of 20 g/litre and precipitate the RNA by adding 2 volumes of ethanol. Cool the solution in ice and leave to stand for 1 h. Collect the precipitate by centrifuging at 2000 g for 5 min in the cold. Wash the RNA with ethanol–water (3:1), ethanol, and, finally, ether; air dry and weigh. (*Note:* Yeast contains about 4 per cent RNA by dry weight.)

Compare your product with a commercial preparation by measuring the pentose, phosphorus and DNA content and determining the absorption spectrum. Keep your preparation for use in later experiments.

Experiment 11.2   Electrophoresis of RNA nucleotides

PRINCIPLE

The ester bonds of RNA are readily hydrolysed due to a free 2′ hydroxyl group on the ribose. This allows the formation of 2′3′-phosphodiester nucleotides which then break down to a mixture of 2′- and 3′-nucleotides. DNA, of course, does not contain a free hydroxyl group on the 2′ position and is therefore relatively stable to dilute alkali. The constituent nucleotides are then separated by electrophoresis in citrate buffer at pH 3.5 where they carry quite different negative charges.

MATERIALS	10
1. Potassium hydroxide (0.3 mol/litre)	100 ml
2. Ribonucleic acid	2 g
3. Perchloric acid (200 g/litre)	100 ml
4. Citrate buffer (0.02 mol/litre, pH 3.5)	5 × tank capacity
5. Hydrochloric acid (0.01 mol/litre)	200 ml
6. Incubator at 37°C	3
7. Ultraviolet spectrophotometer	5
8. Standard nucleotides (AMP, GMP, CMP, and UMP)	—
9. Equipment for cellulose acetate electrophoresis	5
10. Ultraviolet lamp	3

Table 11.1 The absorption of RNA nucleotides in the ultraviolet at pH 2

Nucleotide	Millimolar extinction coefficient at 260 nm E_{260}	E_{250}/E_{260}	E_{280}/E_{260}
Adenylic acid	14.3	0.85	0.22
Guanylic acid	11.8	0.92	0.68
Cytidylic acid	6.8	0.47	1.90
Uridylic acid	9.8	0.78	0.30

Table 11.2 pK values of the ionizable groups on the RNA nucleotides

	Ionizable groups			
	Primary phosphate	Amino	Secondary phosphate	Hydroxyl
Adenylic acid	0.9	3.7	5.9	—
Guanylic acid	0.7	2.3	5.9	9.5
Cytidylic acid	0.8	4.3	6.0	13.2
Uridylic acid	1.0	—	5.9	9.4

METHOD

Dissolve the RNA in 5 ml of 0.3 mol/litre potassium hydroxide (20–30 mg/ml) and incubate at 37°C for 18 h. The following day, place the solution in an ice bath and titrate to pH 3.5 with 200 g/litre perchloric acid. Remove any precipitate by centrifugation and use the supernatant for the electrophoretic separation.

Soak the cellulose acetate in the buffer, gently blot and place across the electrodes. Apply the nucleotides close to the cathode and carry out the electrophoresis at 10 V/cm. The progress of the separation can be readily followed by examining the strip in ultraviolet light. The cellulose acetate fluoresces under ultraviolet irradiation while the nucleotides, which absorb strongly in the ultraviolet, show up as dark spots.

Elute each of the nucleotides with 4 ml of 0.01 mol/litre HCl and plot the absorption spectra of the eluates in the region 220–320 nm. As a blank, use similar eluates from a strip run at the same time with no added nucleotides.

Identify the nucleotides as far as possible from the extinction data given in Table 11.1. Use the pK values of the ionic groups on the nucleotides as given in Table 11.2 to predict their order of separation.

Experiment 11.3 The separation of RNA nucleotides by ion exchange chromatography

PRINCIPLE

As in the previous experiment, RNA is hydrolysed to its constituent nucleotides by dilute alkali. The products of hydrolysis are then separated on a strongly acidic ion exchange column and identified by their characteristic ultraviolet absorption spectra (Table 11.1). UMP and GMP are eluted from the column as distinct peaks but the AMP and CMP are combined and, in this case, an empirical formula is used to calculate the relative amounts of the two nucleotides.

MATERIALS   <u>10</u>
1. Ribonucleic acids from animal and bacterial sources   2 g
2. Potassium hydroxide (0.3 mol/litre)   100 ml
3. Perchloric acid (200 g/litre)   100 ml
4. Hydrochloric acid (0.05 mol/litre)   500 ml
5. Hydrochloric acid (1 mol/litre)   50 ml
6. Dowex 50 W chromatography column (15 cm × 1.5 cm)   5
7. Ultraviolet spectrophotometer   5
8. Volumetric flasks (10 ml)   10
9. Volumetric flasks (25 ml)   5

METHOD

Hydrolyse 0.2 g of RNA with 0.3 mol/litre KOH and neutralize with perchloric acid as described in the previous experiment. Pipette 3.8 ml of the solution into a test tube and add 0.2 ml of 1 mol/litre HCl to give a final concentration of acid of 0.05 mol/litre. Measure the extinction of the solution at 260 nm by diluting a portion with 0.05 mol/litre HCl and apply from 5–12 extinction units to the top of a Dowex 50 W column previously equilibrated with 0.05 mol/litre HCl. Elute with 0.05 mol/litre HCl and collect the effluent in a 10 ml volumetric flask. Monitor the effluent by measuring the extinction at 260 nm and, when the first nucleotide has emerged from the column, place a second 10 ml flask containing 0.5 ml of 1 mol/litre HCl under the column and elute with water until all the GMP has been eluted. At this stage increase the flow rate down the column and collect the AMP and CMP into a 25 ml flask containing 1.25 ml of 1 mol/litre HCl.

Make up the first flask with 0.05 mol/litre HCl and the other two flasks with water. Mix thoroughly and plot the ultraviolet spectra of a sample from each flask from 220 nm to 320 nm. Explain why the nucleotides are eluted in the above order and use the formulae below to calculate the amount of each nucleotide present. Does the simple relationship AMP/UMP = GMP/CMP = 1 hold for RNA?

Peak 1: UMP(μmoles) = $(E_{260} \times vol)/9.8 = (E_{260} \times 10)/9.8$

Peak 2: GMP(μmoles) = $(E_{257} \times vol)/11.8 = (E_{257} \times 10)/11.8$

Peak 3:   If $y = (2.32E_{257} - E_{279})/2.08$

$$CMP(\mu moles) = (E_{279} - 0.238y) \times vol/13.2$$

$$= (E_{279} - 0.238y) \times 25/13.2$$

$$AMP(\mu moles) = (y \times vol)/14.3 = (y \times 25)/14.3$$

Experiment 11.4   The base composition of RNA

PRINCIPLE

Ribonucleic acid and deoxyribonucleic acid can be hydrolysed to the constituent bases by treatment with 72 per cent perchloric acid for 1 h. The method is not completely quantitative since some thymine is lost. The resulting bases are then separated by paper chromatography and detected with ultraviolet light.

MATERIALS	10
1. Ribonucleic acid	1 g
2. Perchloric acid (72 per cent)	20 ml
3. Marker bases (adenine, guanine, cytosine, uracil, and thymine, 5 mmol/litre)	1 ml
4. Boiling water bath	5
5. Ultraviolet lamp	5
6. Hydrochloric acid (0.1 mol/litre)	250 ml
7. Ultraviolet spectrophotometer	5
8. Hydrochloric acid (conc.)	—
9. Isopropanol	—
10. Paper chromatography equipment	5
11. Chromatography solvent (isopropanol:water:conc. HCl, 130:37:33)	5 × tank capacity

METHOD

Mix the nucleic acid (100 mg) with 1 ml of perchloric acid and heat on a boiling water bath for 1 h. Place a marble or ampoule on top of the tube to reduce the loss by evaporation. *This hydrolysis must be carried out in a fume chamber behind a protective screen since there is a risk of an explosion. Do not heat to dryness.*

Cool the tube, add 1 ml of water, and centrifuge the contents: use the clear supernatant for descending chromatography overnight on Whatman No. 1 paper. Marker spots of the bases should be run at the same time.

Dry the chromatogram in a current of cold air for 4 h, then locate the bases with ultraviolet light. Elute the spots with 5 ml of 0.1 mol/litre HCl overnight and plot the

Table 11.3 The absorption of purine and pyrimidine bases in 0.1 mol/litre hydrochloric acid

Nucleotide bases	Wavelength maximum	Millimolar extinction coefficient
Adenine	260	13.0
Guanine	250	11.2
Uracil	260	7.9
Thymine	265	7.9
Cytosine	275	10.5

absorption spectra. Identify the bases by comparing the absorption spectra with standard purines and pyrimidines and determine the relative amounts present in the nucleic acid from the absorption coefficients given in Table 11.3.

Compare the base ratios with those obtained by alkaline hydrolysis followed by electrophoresis of the nucleotides.

Experiment 11.5   The isolation of DNA from pig spleen

PRINCIPLE

Almost all cells contain DNA, but the amount present in some tissues is quite small so that they are not a particularly convenient source. In addition, some tissues contain high deoxyribonuclease (DNase) activity so that the DNA is broken down into smaller fragments. A convenient source for the isolation of DNA should therefore contain a high quantity of the material and have low deoxyribonuclease activity. Lymphoid tissue is very good in these respects and thymus is the best source, with spleen as a good alternative.

DNA is readily denatured and extreme care must be taken in order to obtain a product that is structurally related to that found in the cell. Mechanical stress and extreme physical and chemical conditions must be avoided and nucleases must be inhibited. Sodium citrate is therefore present in the solution to bind Ca^{2+} and Mg^{2+} which are cofactors for DNase and the stages until the removal of protein are carried out as rapidly as possible in the cold.

The nucleoprotein is soluble in water and solutions of high ionic strength, but is insoluble in solutions of low ionic strength (0.05–0.25 mol/litre) and use is made of this property in the initial extraction. The tissue is first homogenized in isotonic saline buffered with sodium citrate, pH 7, when most other macromolecules pass into solution, leaving the insoluble deoxyribonucleoprotein which is then dissolved in 2 mol/litre saline.

The protein is removed by treatment with a chloroform/amyl alcohol mixture and the DNA precipitated with ethanol. The product is then dissolved in dilute buffered saline and stored frozen. It is stable in this form for several months.

MATERIALS   10

1. Pig spleen   300 g
2. Buffered saline (0.15 mol/litre NaCl buffered with 0.015 mol/litre   3 litres
   sodium citrate, pH 7)
3. Sodium chloride (2 mol/litre)   6 litres
4. Chloroform: amyl alcohol (6:1)   2 litres
5. Absolute ethanol   2 litres
6. Ether   500 ml
7. Waring blender   5

METHOD

Chop 50 g of pig spleen into small fragments and homogenize with 200 ml of buffered saline for 1 min. Centrifuge the suspension at 5000 g for 15 min and rehomogenize the precipitate in a further 200 ml of buffered saline. Discard the supernatant and suspend the combined sediments uniformly in 2 mol/litre NaCl to a final volume of 1 litre, when most of the material should dissolve. Remove any sediment by centrifugation and stir the solution continuously with a glass rod while adding an equal volume of distilled water. Spool the fibrous precipitate on to a glass rod and leave it to stand in a beaker for 30 min. During this time the clot will shrink and the liquid expressed should be removed with filter paper.

Dissolve the deoxyribonucleoprotein in about 100 ml of 2 mol/litre NaCl, add an equal volume of the chloroform/amyl alcohol mixture (6:1), and blend for 30 s. Centrifuge the emulsion at 5000 g for 10–15 min and collect the upper (opalescent) aqueous layer containing the DNA. This is best carried out by gentle suction into a suitable container so that the denatured protein at the interface of the two liquids is not disturbed. Repeat the treatment with organic solvent twice more and collect the supernatant in a 500 ml beaker.

Precipitate the DNA by slowly stirring 2 volumes of ice-cold ethanol with the supernatant and collect the mass of fibres on the glass stirring rod. Carefully remove the rod and gently press the fibrous DNA against the side of the beaker to expel the solvent. Finally, wash the precipitate by dipping the rod into a series of solvents and expelling the solvent as described. Four solvents are used: 70 per cent v/v ethanol, 80 per cent v/v ethanol, absolute ethanol and ether. Remove the last traces of ether by standing the DNA in a fume cupboard for about 10 min.

Weigh the dry DNA and dissolve by continuously stirring in buffered saline diluted one in ten with distilled water (2 mg/ml); store frozen until required.

Experiment 11.6   The ultraviolet absorption of the nucleic acids

PRINCIPLE

The nucleic acids absorb strongly in the ultraviolet region of the spectrum due to the conjugated double bond systems of the constituent purines and pyrimidines. They show characteristic maxima at 260 nm and minima at 230 nm.

The water content of nucleic acids is not usually known, so extinction coefficients cannot be reliably based on weight. The extinction is conveniently expressed as E_P, the extinction of a solution containing 1 mol/litre of phosphorus. Even so, the extinction coefficient of a nucleic acid is not a constant but depends on the previous treatment of the material, as well as the pH and ionic strength of the medium. Typical values for E_P are given below:

$$E_P \text{ for DNA} = 6000\text{--}8000$$

$$E_P \text{ for RNA} = 7000\text{--}10\,000$$

The extinction coefficient of a nucleic acid may increase by up to 40 per cent on degradation or hydrolysis and this is known as the *hyperchromic effect*. In the macromolecule, hydrogen bonding and π–π interactions alter the resonance behaviour of the bases so that the extinction of the nucleic acid is less than that of the constituent nucleotides.

When a solution of double-stranded DNA is slowly heated, there is little change in extinction until the 'melting temperature' (T_m) is reached; at this stage, the absorbance increases rapidly to a higher value, which is not significantly changed by further heating. At the melting temperature, the hydrogen bonds between base pairs on opposite strands are broken and the two DNA threads are separated. If the hot DNA solution is then cooled slowly, the two threads recombine and the 'cooling curve' should be superimposed on the 'melting curve'. If, however, the DNA is cooled rapidly, then some recombination of the two strands takes place, but in a more random manner so that the extinction of the solution at room temperature is higher than that of the original DNA solution before heating (Fig. 11.6).

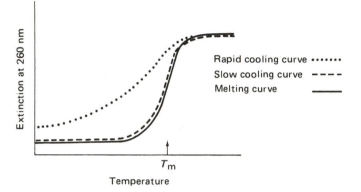

Fig. 11.6   The effect of temperature on the extinction of DNA

MATERIALS   10
1. Deoxyribonucleic acid   0.5 g
2. Ribonucleic acid   0.5 g
3. Deoxyribonuclease (DNase)   2 ml
4. Acetate buffer (0.5 mol/litre, pH 5.5)   1 litre
5. Acetate buffer–magnesium sulphate solution. (Prepare a   1 litre
   0.1 mol/litre solution of magnesium sulphate in the above buffer
   solution)
6. Buffered saline (0.15 mol/litre NaCl, 0.105 mol/litre sodium citrate,   1 litre
   pH 7)
7. Ultraviolet spectrophotometer with a thermostatically controlled   5
   cell housing
8. Water bath for use at temperatures up to 95°C   5

METHOD

Spectrophotometric assay   Dissolve 10 mg of RNA and DNA in the buffered saline and make up to 100 ml. Plot the absorption spectra of these two solutions between

220 and 320 nm and express the final results as E_P after determining the phosphorus content as in Exp. 11.10.

Compare the results obtained for your own preparations with those for commercial samples and comment on any differences observed.

The action of deoxyribonuclease (deoxyribonucleate oligonucleotidehydrolase, 3.1.4.4.) Prepare a solution of DNA in the acetate buffer–magnesium sulphate solution (10 mg/100 ml) and pipette 2.5 ml of this into a cuvette. Add 0.5 ml of enzyme solution, suitably diluted, and follow the change in extinction at 260 nm. Compare this with a blank containing distilled water instead of the enzyme.

Repeat the experiment with RNA and explain the results.

The effect of heat on DNA Prepare three solutions of DNA in dilute (0.1), standard (1), and strong ($\times$ 5) buffered saline to give an initial extinction value of about 0.3–0.5 (0.1 mg/ml). Place the solutions in three cuvettes in a spectrophotometer with a thermostatically controlled cell housing and record the change in extinction as the temperature is increased to 90°C. The temperature of the water bath heating the cell container is usually higher than that of the contents of the cuvettes, so the temperature of the DNA solutions should be measured directly if possible.

In addition, measure the extinction change as the cuvettes are slowly cooled from 90°C to room temperature. Repeat the experiment, but this time cool the solutions rapidly.

Compare the cooling and melting curves for the two experiments and comment on the effect of ionic strength on the melting temperature.

Experiment 11.7   The viscosity of DNA solutions

PRINCIPLE

The separation of the complementary strands of DNA involves a change from the rigid linear double helix to a random coil formation and leads to a drop in viscosity. When using the Ostwald viscometer (Fig. 8.1), slow rates of flow must be used and very dilute solutions to make sure that the molecules are randomly orientated and not aligned by fluid flow.

The relative viscosity is obtained from viscometer timings of the test solution (t_1) and the solvent (t_0) so that:

$$\text{Relative viscosity } \eta_{rel} = t_1/t_0$$

$$\text{Specific viscosity } \eta_{sp} = \eta_{rel} - 1 = (t_1 - t_0)/t_0$$

$$\text{Reduced viscosity } \eta_{red} = (t_1 - t_0)/ct_0$$

$$\text{where } c = \text{DNA concentration in mg/ml.}$$

The *intrinsic viscosity η* is the viscosity at infinite dilution and is obtained by plotting η_{red} against concentration and reading the intercept when $c = 0$. The intrinsic viscosity depends on the properties of the molecule and is related to the mol. wt of the DNA so that:

$$0.665 \log M = 3.86 + \log (\eta + 0.5)$$

MATERIALS $\underline{10}$
1. Ostwald viscometers and 25°C water baths   5
2. Buffered saline (0.15 mol/litre NaCl, 0.015 mol/litre sodium citrate,   1 litre
   pH 7)
3. Bacterial DNA or calf thymus DNA   100 mg
4. Stock DNA solution (1 mg/ml in buffered saline)   100 ml

METHOD
The details for the operation of the Ostwald viscometer and the determination of relative viscosity are given in Exp. 8.9.

Determination of mol. wt Prepare a range of concentrations of DNA in the buffered saline (0, 0.05, 0.1, 0.2, 0.4 mg/ml). Equilibrate in the water bath at 25°C and measure the flow times of each of the solutions. Determine the intrinsic viscosity and the mol. wt of the DNA.

The effect of denaturation Take the highest concentration of DNA (0.4 mg/ml) and boil for 5 min; cool to 25°C and measure the viscosity. Explain the results obtained.

Hydrolysis by DNase Investigate the effects of deoxyribonuclease on the viscosity of the DNA using the conditions given in Exp. 11.6.

Experiment 11.8   The estimation of DNA by the diphenylamine reaction

PRINCIPLE
When DNA is treated with diphenylamine under acid conditions, a blue compound is formed with a sharp absorption maximum at 595 nm. This reaction is given by 2-deoxypentoses in general and is not specific for DNA. In acid solution, the straight chain form of a deoxypentose is converted to the highly reactive β-hydroxylevulinaldehyde which reacts with diphenylamine to give a blue complex. In DNA, only the deoxyribose of the purine nucleotides reacts, so that the value obtained represents half of the total deoxyribose present.

MATERIALS $\underline{10}$
1. DNA (commercial sample)   10 mg
2. RNA (commercial sample)   10 mg
3. Pig spleen DNA solution from Exp. 11.5   10 mg
4. Yeast RNA solution from Exp. 11.1   10 mg
5. Buffered saline (0.15 mol/litre NaCl; 0.015 mol/litre sodium citrate, pH 500 ml
   7)

6. Diphenylamine reagent. (Dissolve 10 g of pure diphenylamine in 1 litre   1 litre
   of glacial acetic acid and add 25 ml of concentrated sulphuric acid. This
   solution must be prepared fresh)
7. Boiling water bath   5

METHOD

Dissolve 10 mg of the nucleic acid in 50 ml of buffered saline, remove 2 ml and add 4 ml of diphenylamine reagent. Heat on a boiling water bath for 10 min, cool and read the extinction at 595 nm. Read the test and standards against a water blank. Assay the isolated nucleic acids and the commercial samples for DNA.

Experiment 11.9   The estimation of RNA by means of the orcinol reaction

PRINCIPLE

This is a general reaction for pentoses and depends on the formation of furfural when the pentose is heated with concentrated hydrochloric acid. Orcinol reacts with the furfural in the presence of ferric chloride as a catalyst to give a green colour. Only the purine nucleotides give any significant reaction.

MATERIALS   10
1. DNA and RNA solutions prepared as in Exp. 11.8 (0.2 mg/ml)   250 ml
2. Orcinol reagent. (Dissolve 1 g of ferric chloride ($FeCl_3 \cdot 6H_2O$) in   1 litre
   1 litre of concentrated HCl and add 35 ml of 6 per cent w/v orcinol
   in alcohol)
3. Boiling water bath   5

METHOD

Mix 2 ml of the nucleic acid solution with 3 ml of orcinol reagent. Heat on a boiling water bath for 20 min, cool and determine the extinction at 665 nm against an orcinol blank.

Experiment 11.10   The determination of the phosphorus content of a nucleic acid

PRINCIPLE

The nucleic acid is oxidized with perchloric acid to give inorganic phosphate which is then estimated by the usual methods.

MATERIALS   10
1. Reagents for the estimation of inorganic phosphate (Exp. 7.3)   —
2. Perchloric acid (60 per cent v/v)   10 ml
3. Digestion flasks and racks   20

METHOD

Pipette 2 ml of a solution of nucleic acid (0.1 mg/ml) into a digestion flask. Add 0.5 ml of perchloric acid and digest over a low flame for 1 h until all the inorganic matter has disappeared. (*Care:* explosion risk, see Exp. 11.4.) After cooling, add 1 ml of distilled water to each flask and measure the phosphate content as described in Exp. 7.3.

Further reading

Adams, R. C. P., Burdon, R. H., Campbell, A. M., Leader, D. P. and Smellie, R. M. S., *The Biochemistry of the Nucleic Acids*, 9th edn. Chapman and Hall, London, 1981.

Mainwaring, W. I. P., Parish, J. H., Pickering, I. D. and Mann, N. H., *Nucleic Acid Biochemistry and Molecular Biology*. Blackwell Scientific Publications, Oxford, 1982.

Watson, J. D., *The Double Helix*. Weidenfeld and Nicolson, London, 1968.

SECTION FOUR Metabolic studies

12. Enzymes

Enzymes as catalysts

Catalysis

Living organisms are able to obtain and use energy very rapidly because of the presence of biological catalysts called enzymes. As with inorganic catalysts, enzymes change the rate of a chemical reaction but do not affect the final equilibrium; also, only small quantities are needed to bring about the transformation of a large number of molecules. However, unlike most inorganic catalysts, enzymes have a very narrow specificity, i.e., they will only catalyse a comparitively small range of reactions or, in some cases, only one reaction. Enzymes will also only function under certain well-defined conditions of pH, temperature, substrate concentration, cofactors, etc., and these properties are illustrated in some of the following experiments.

Classification

Enzymes are named and classified according to the type of reaction catalysed. The six main groups of enzymes are:

1. Oxidoreductases
2. Transferases
3. Hydrolases
4. Lyases
5. Isomerases
6. Ligases (synthetases)

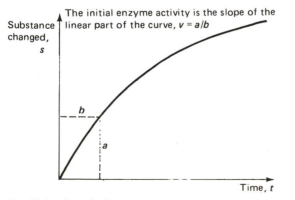

Fig. 12.1 A typical progress curve

Each enzyme has a systematic name and number which identifies it by indicating its group and subgroups. Lactate:NAD oxidoreductase, for example, is number 1.1.1.27. The first number tells us that the enzyme is in group 1 and therefore an oxidoreductase. The second number is the subgroup and signifies the chemical group changed, which in this case is —CHOH. The third number denotes the subsubgroup, which shows that NAD or NADP is the hydrogen acceptor. The fourth and final figure is the characteristic number of the enzyme. There is also a recommended trivial name which is shorter and more convenient than the systematic name and the above enzyme is more commonly known as lactate dehydrogenase.

Measuring enzyme activity

Enzyme assay

Enzymes are assayed by following the disappearance of the substrate or the appearance of the reaction products with time.

Controls

In addition to the 'test' mixture, 'controls' are always prepared, one containing no enzyme and the other containing the enzyme but no substrate. Other control mixtures may be required when the enzyme has several cofactors. The object of these controls is to allow for non-specific and spontaneous chemical reactions not catalysed by the enzyme.

Progress curve

A plot of the amount of substrate changed or product formed with time is known as a *progress curve* (Fig. 12.1). This is linear at first but then falls as the reaction proceeds.

The *enzyme activity* (v) is obtained from the linear part of the curve which is the *initial reaction velocity* ($v = a/b$).

Enzyme units

The *enzyme activity* is most frequently expressed in terms of units (U) such that *one unit is the amount of enzyme that catalyses the conversion of 1 micromole of substrate per minute under defined conditions*. In some cases the unit is too large and the activity can be more conveniently expressed in terms of nmol/min or pmol/min.

The SI unit of enzyme activity is the *katal (kat) which represents the transformation of 1 mole of substrate per second*. This unit is big and more manageable figures are obtained by expressing activities in microkatals (μkat), nanokatals (nkat) or picokatals (pkat).

$$1 \text{ U} = 1 \text{ }\mu\text{mol min}^{-1}$$

$$1 \text{ katal} = 1 \text{ mol s}^{-1}$$

$$1 \text{ U} = \mu\text{kat}/60 = 16.67 \text{ nkat}$$

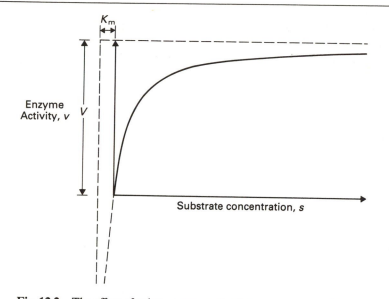

Fig. 12.2   The effect of substrate concentration on enzyme activity

The purity of an enzyme is expressed in terms of the *specific activity*, which is the number of enzyme units (U) per milligram of protein. The specific activity in SI units is given as katals per kilogram of protein.

Enzyme activity and substrate concentration

Michaelis–Menten enzymes

Kinetics   If the activity of an enzyme is determined over a range of substrate concentrations a curve similar to the rectangular hyperbola of Fig. 12.2 is often obtained.

At low substrate concentrations v varies linearly with s, giving *first-order kinetics*: $v = -ds/dt = ks$, where k is the rate constant.

At high substrate concentrations v is independent of s, giving *zero-order kinetics*: $v = -ds/dt = \text{constant} = V$.

At intermediate substrate concentrations there is a mixture of first- and zero-order kinetics.

The Michaelis equation   An equation relating v and s can be obtained for the whole curve and was first derived by Michaelis and Menten in 1913:

$$v = \frac{V}{1 + K_m/s} \quad \text{or} \quad v = \frac{Vs}{s + K_m} \tag{1}$$

where V and K_m are constants.

This equation can be rearranged to give the form usually shown for the equation of a rectangular hyperbola (Fig. 12.2).

$$(V-v)(K_\mathrm{m}+s) = VK_\mathrm{m}$$

The basic assumption used by these authors was that the enzyme and substrate form a complex which then breaks down to give the enzyme and products.

The whole process can be represented by the general equation:

$$\mathrm{E} \; + \; \mathrm{S} \underset{k_{-1}}{\overset{k_{+1}}{\rightleftharpoons}} \mathrm{ES} \xrightarrow{k_{+2}} \mathrm{E} + \mathrm{Products}$$

$$(e-p) \quad (s) \qquad\quad (p)$$

where e = enzyme (E) concentration, s = substrate (S) concentration, p = enzyme–substrate complex (ES) concentration, k_{+1}, k_{+2}, k_{-1} are velocity constants.

The kinetic constants K_m *and* V   If $s = K_\mathrm{m}$, then $v = V/2$ so that the Michaelis constant is the substrate concentration which gives half the maximum velocity.

If the equilibrium conditions suggested by Michaelis and Menten hold, then k_{+2} is small compared with k_{+1} and can be ignored, so that $K_\mathrm{m} = k_{-1}/k_{+1}$, which is the dissociation constant for the enzyme–substrate complex. A large K_m therefore means a large dissociation constant or a small association constant ($1/K_\mathrm{m}$). Conversely a small K_m means a small dissociation constant or a large association constant ($1/K_\mathrm{m}$). The Michaelis constant therefore gives a measure of the enzyme–substrate affinity.

Large K_m = low enzyme–substrate affinity.

Small K_m = high enzyme–substrate affinity.

The kinetic constants K_m and V are most conveniently determined from a linear transformation of the Michaelis equation, obtained by taking reciprocals:

$$\frac{1}{v} = \frac{1}{V} + \frac{K_\mathrm{m}}{V} \times \frac{1}{s}$$

A plot of $1/v$ against $1/s$ therefore gives a straight line of slope K_m/V (Fig. 12.3(a)). The reciprocals of the kinetic constants can then be determined from the intercepts on the axis, since when

$$\frac{1}{s} = 0, \quad \frac{1}{v} = \frac{1}{V}$$

and when

$$\frac{1}{v} = 0, \quad \frac{1}{s} = \frac{1}{K_\mathrm{m}}$$

An alternative plot is of s/v against s (Fig. 12.3(b)). This is obtained by multiplying the reciprocal equation by s:

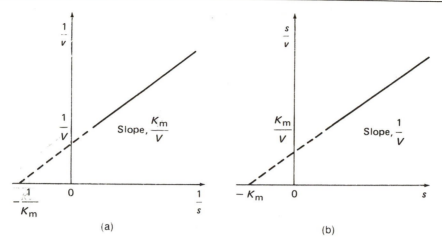

Fig. 12.3   The determination of the Michaelis constant

$$\frac{s}{v} = \frac{s}{V} + \frac{K_m}{V}$$

when

$$s = 0, \quad \frac{s}{v} = \frac{K_m}{V}$$

and

$$\frac{s}{v} = 0, \quad s = -K_m$$

The graph of $(1/v)$ against $(1/s)$ is known as the Lineweaver–Burk plot and is the method most frequently used to calculate K_m. However, the slope of the line is most influenced by the activities measured at low substrate concentrations, which are the least accurate, and, for this reason, the plot of (s/v) against s is to be preferred.

Validity of the Michaelis equation   Most enzymes give a rectangular hyperbola when v is plotted against s in support of the Michaelis–Menten theory. There are some exceptions to this and these are dealt with in the next section on allosteric enzymes.

Allosteric enzymes

Effect of substrate   There are some enzymes that give a sigmoid curve when the activity is plotted against substrate concentration (Fig. 12.4) and do not obey the usual Michaelis–Menten kinetics. These enzymes are made up of subunits and contain more than one active site per molecule. The sigmoid curve of v versus s can be explained on the basis that when one molecule binds, a *conformational change* takes place in the

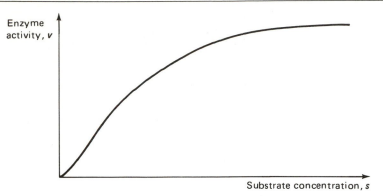

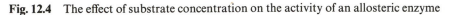

Fig. 12.4   The effect of substrate concentration on the activity of an allosteric enzyme

molecule enabling the next molecule of substrate to bind more readily. The rate of substrate binding therefore depends on the number of active sites already occupied by substrate molecules. Haemoglobin, although not an enzyme, is a good example of an allosteric protein with four binding sites for oxygen. The relative ease of binding of the oxygen atoms from the first to the fourth is approximately in the ratio of 1:4:24:9.

Allosteric kinetics   The derivation of an accurate equation for the sigmoid curve of allosteric enzymes is complicated, but a simplified equation can be obtained, as suggested by Atkinson, if the following assumptions are made:

1. The intermediates ES, ES_2, ES_3, etc., have only a transient existence.
2. Equilibrium is rapidly attained between the enzyme and the substrate molecules.

If there are n binding sites for the substrate S, then:

$$E + nS \underset{k_{-1}}{\overset{k_{+1}}{\rightleftharpoons}} ES_n \xrightarrow{k_{+2}} E + \text{Products}$$

$$\frac{v}{V-v} = \frac{[S]^n}{K} \qquad (2)$$

Rearranging the equation,

$$v = \frac{V[S]^n}{K + [S]^n} \qquad (3)$$

If there is only one binding site for the substrate ($n = 1$) then this equation becomes the Michaelis equation and a hyperbolic plot is obtained.

Kinetic constants   A convenient linear transformation can be obtained by taking logarithms of Eq. (2). The kinetic constants can then be read off the straight-line plot (Fig. 12.5).

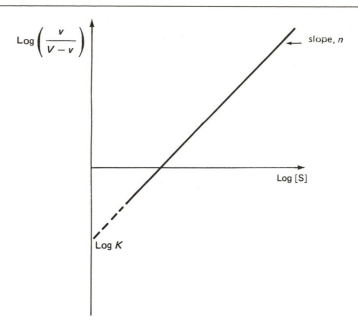

Fig. 12.5 The determination of the kinetic constants for an allosteric enzyme

$$\log v/(V-v) = n\log[\mathrm{S}] - \log K$$

n is a measure of the number of binding sites if the original assumptions are correct, but in practice n is usually less than this since the existence of the intermediates is not transient. n is the *degree of cooperativity*: the greater the value of n, the more sigmoid the plot of v versus $[\mathrm{S}]$.

K is actually a complex steady-state constant and for allosteric enzymes, when $[\mathrm{S}] = K$, v is *not* $V/2$, but if $v = V/2$ then, from Eq. (2):

$$2[\mathrm{S}]^n = K$$

The substrate concentration which gives 50 per cent of the maximum activity is thus related to the constant K so that:

$$K = [\mathrm{S}]_{50}^n$$

or. taking logs,

$$\log K = n\log[\mathrm{S}]_{50}$$

Metabolic control   Allosteric enzymes have specific sites for the binding of activators and inhibitors which act by causing a change in the conformation of the protein. This enables compounds that are chemically unrelated to the substrate to affect the enzyme activity, and allosteric enzymes are found at many points of metabolic control.

Factors affecting enzyme activity

Cofactors

In many cases, if an enzyme is mixed with its substrate under the appropriate conditions, either no catalysis occurs or there is only a slight activity. This is often due to the absence of a coenzyme or activator.

Coenzymes   These are low molecular weight organic compounds which are actively involved in the catalysis. They often act as acceptors or donors of specific chemical groups. NAD, for example, accepts and donates hydrogen atoms and is a coenzyme for many dehydrogenases. The name coenzyme is reserved for soluble cofactors while the term *prosthetic group* is used for coenzymes that are firmly attached to the protein.

Activators   These are of a simple chemical nature and are not so specific as coenzymes. They appear to function by activating the enzyme–substrate complex. Several metal ions are known to be activators of a wide range of enzymes; Mg^{2+} for alkaline phosphatase and the kinases.

Inhibitors

Many compounds react with enzymes and reduce the measured activity. This property of enzymes is used in designing drugs and insecticides which selectively inhibit enzymes in the infective bacteria or insects, but do not affect the animal or plant. Two classical types of inhibition are recognized: competitive and non-competitive.

Competitive inhibition   In this case, the inhibitor reacts with the enzyme by competing with the substrate for the active site. The degree of inhibition depends on the relative concentrations of substrate and inhibitor, and almost maximal velocity may be found in the presence of the inhibitor if the substrate concentration is high enough.

Competitive inhibitors are of a similar chemical structure to the natural substrate and are fairly specific. This is well illustrated in the case of the enzyme succinate dehydrogenase, which catalyses the conversion of succinate to fumarate. Malonate and maleate both act as competitive inhibitors of this enzyme.

$$
\begin{array}{lll}
 & \underset{\text{Succinate}}{\overset{\displaystyle CH_2COOH}{\underset{\displaystyle CH_2COOH}{|}}} + FAD \rightleftharpoons & \underset{\text{Fumarate}}{\overset{\displaystyle CHCOOH}{\underset{\displaystyle HOOCCH}{\|}}} + FADH_2
\end{array}
$$

Substrates: Succinate   Fumarate

$$
\begin{array}{ll}
\underset{\text{Malonate}}{\overset{\displaystyle COOH}{\underset{\displaystyle COOH}{\overset{\displaystyle |}{\underset{\displaystyle |}{CH_2}}}}} & \underset{\text{Maleate}}{\overset{\displaystyle CH\cdot COOH}{\underset{\displaystyle CH\cdot COOH}{\|}}}
\end{array}
$$

Inhibitors: Malonate   Maleate

Non-competitive inhibition   In the non-competitive type of inhibition, the inhibitor combines with the enzyme but not at the active site, so that the enzyme can bind both

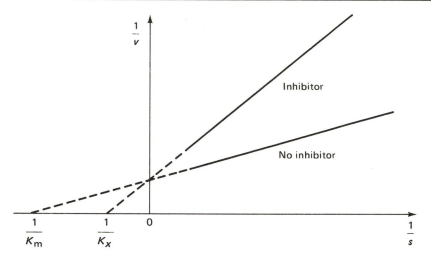

Fig. 12.6   The effect of a competitive inhibitor on the Lineweaver–Burk plot

substrate and inhibitor at the same time. The binding site of the inhibitor is usually sufficiently far removed from the active centre so that the binding of the substrate is unaffected. The enzyme–substrate–inhibitor complex formed is unable to break down and inhibition effectively occurs by the reduction of the amount of enzyme available. Increase of the substrate concentration has no effect on the degree of inhibition.

Most non-competitive inhibitors are not related chemically to the substrate and the same inhibitor may affect a number of enzymes.

Examples of non-competitive inhibition are the action of thiol blocking agents such

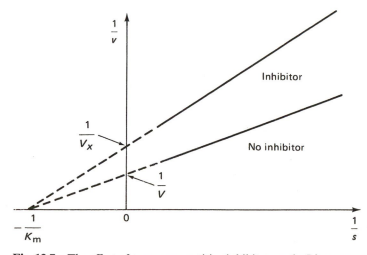

Fig. 12.7   The effect of a non-competitive inhibitor on the Lineweaver–Burk plot

as *p*-chloromercuribenzoate, heavy metal ions such as Ag^+ and Cu^{2+}, and the reaction of cyanide with the iron–porphyrin enzymes.

Determination of inhibitor constant (K_i)   In the presence of an inhibitor, the Michaelis–Menten equation is altered as shown below, where i is the concentration of the inhibitor and K_i the inhibition constant:

No inhibitor,
$$v = \frac{V}{1 + (K_m/s)}$$

Competitive inhibitor,
$$v = \frac{V}{1 + K_m/s(1 + i/K_i)}$$

Non-competitive inhibitor,
$$v = \frac{V}{(1 + K_m/s)(1 + i/K_i)}$$

Taking reciprocals of the above equations we obtain:

No inhibitor,
$$\frac{1}{v} = \frac{1}{V} + \frac{K_m}{V}\frac{1}{s}$$

Competitive inhibitor,
$$\frac{1}{v} = \frac{1}{V} + \frac{K_m}{V}\left(1 + \frac{i}{K_i}\right)\frac{1}{s}$$

Non-competitive inhibitor,
$$\frac{1}{v} = \frac{1}{V}\left(1 + \frac{i}{K_i}\right) + \frac{K_m}{V}\left(1 + \frac{i}{K_i}\right)\frac{1}{s}$$

A graph of $(1/v)$ against $(1/s)$ can be used to determine K_i and the type of inhibition as shown below (Figs. 12.6 and 12.7). A competitive inhibitor alters K_m while the non-competitive type causes a change in V.

$$K_x = K_m\left(1 + \frac{i}{K_i}\right)$$

$$V_x = \frac{V}{(1 + i/K_i)}$$

Temperature

The effect on the enzyme reaction   Molecules must possess a certain *energy of activation* (E) before they can react, and enzymes function as catalysts by lowering this energy of activation, thereby enabling the reaction to proceed more rapidly. The overall change in the free energy (ΔG) is unaffected by the enzyme.

The energy of activation of an enzyme-catalysed reaction can be determined by measuring the maximum velocity (V) at different temperatures, and plotting $\log_{10} V$ against $1/T$. The slope of the line is given by $-E/2.303\,R$. This relationship is obtained from the empirical equation of Arrhenius:

$$\mathrm{d}\ln k/\mathrm{d}T = E/RT^2$$

Integrating this equation one obtains:

$$\log_{10} k = C - E/2.303RT$$

where

C = a constant,
k = velocity constant of reaction,
T = temperature (K),
R = gas constant = $8.32\ \text{J mol}^{-1}\ \text{K}^{-1}$,
E = energy of activation (J mol^{-1}).

The velocity constant is not easy to obtain and since V is directly proportional to k, the maximum velocity is plotted against the reciprocal of the temperature for the Arrhenius plot (Fig. 12.8).

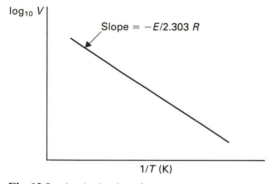

Fig. 12.8 An Arrhenius plot

The effect of temperature on denaturation This can be determined by exposing the enzyme to a high temperature for different periods of time, then measuring the activity at a temperature at which the enzyme is stable. This will tell us how much of the enzyme has been destroyed and a graph can be prepared of the rate of loss of active enzyme with time. This is repeated for a number of different temperatures and the initial rate plotted against $1/T$. The value of E from this plot is the *energy of inactivation* which is usually quite high due to the large positive *entropy* change resulting from the unfolding of the molecule during denaturation.

pH

The pH optimum Enzymes are active over a limited pH range only and a plot of activity against pH usually gives a bell-shaped curve of the type shown in Fig. 12.9. The pH value of maximum activity is known as the optimum pH and this is a characteristic of the enzyme, provided that the enzyme is stable under the conditions studied.

The variation of activity with pH is due to the change in the state of ionization of the

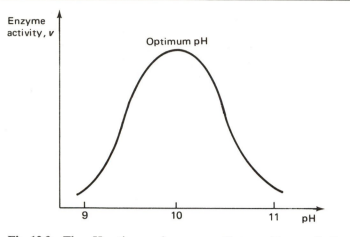

Fig. 12.9   The pH optimum of an enzyme illustrated by rat alkaline phosphatase

enzyme protein and other components of the reaction mixture. Michaelis and Davidsohn suggested in 1911 that only one of the large number of ionized forms of the protein is active, so that a change in pH either side of the optimum produces a decrease of this form and hence a fall in the activity.

K_m *and* V   Changing the pH alters the enzyme activity by affecting V, K_m, or the stability of the enzyme protein. Most published plots of pH optima are the result of changes in V and K_m and are not, therefore, readily amenable to mathematical treatment. Ideally the effect of pH on K_m and V should be studied separately.

Enzyme stability   If the enzyme is unstable at certain pH values, then the optimum pH is no longer a characteristic of the enzyme. The stability can be checked by exposing the enzyme to the appropriate pH value for the time of the experiment and then adjusting the pH to a value at which the enzyme is stable and measuring the activity.

Experiments

Experiment 12.1   The progress curve obtained during the hydrolysis of *p*-nitrophenyl phosphate by serum alkaline phosphatase (orthophosphoric monoester phosphohydrolase, 3. 1. 3. 1)

PRINCIPLE
Alkaline phosphatase (pH 9–10 optimum) is found in bone, kidney, liver and intestine and acts on phosphoric esters with the liberation of inorganic phosphate.

$$R \cdot O \cdot \overset{\displaystyle OH}{\underset{\displaystyle OH}{P}}{=}O \;+\; H_2O \;\longrightarrow\; R \cdot OH \;+\; HO{-}\overset{\displaystyle OH}{\underset{\displaystyle OH}{P}}{=}O$$

p-Nitrophenyl phosphate is used as substrate and the p-nitrophenol released by enzymic hydrolysis is measured colorimetrically. In alkaline solution, p-nitrophenol absorbs at 405 nm. The substrate p-nitrophenyl phosphate does not absorb at this wavelength so the progress of the enzyme-catalysed reaction can be readily followed by measuring the change in extinction at 405 nm.

MATERIALS   100
1. Sodium carbonate–bicarbonate buffer (0.1 mol/litre, pH 10.0)   500 ml
2. p-Nitrophenyl phosphate substrate solution. (5 mmol/litre in the   500 ml
   alkaline buffer)
3. p-Nitrophenol standard (50 μmol/litre). This is prepared by   1 litre
   dissolving 69.6 mg p-nitrophenol in 100 ml of alkaline buffer, then
   diluting this solution one in a hundred with the buffer. Solutions (2)
   and (3) should be freshly prepared
4. Serum   50 ml
5. Colorimeters   50

METHOD

Test   Add 0.8 ml of the carbonate–bicarbonate buffer to 2 ml of the substrate solution and mix thoroughly. At zero time, add 0.2 ml of serum and follow the change in extinction at 405 nm.

Blank   This is the same as the test except that 0.2 ml of buffer replaces the serum.

Standard   Prepare a range of p-nitrophenol solutions and plot a graph of extinction at 405 nm against concentration.

CALCULATION
Subtract the extinction of the blank from that of the test and calculate the amount of p-nitrophenol released from the standard curve. Express the activity of the serum alkaline phosphatase in terms of units per millilitre of serum remembering that 0.2 ml is used in the assay.

Experiment 12.2   Variation of serum alkaline phosphatase activity with enzyme concentration

PRINCIPLE
A plot of enzyme activity against enzyme concentration should give a straight line unless an inhibitor or activator is present in the sample or the activity is wrongly calculated from the non-linear part of the progress curve.

MATERIALS
Prepare 2 litres of buffer and substrate and 200 ml of serum as in the previous experiment. This is sufficient material for 50 pairs of students.

METHOD

Repeat the previous experiment with volumes of serum from 0.1 to 0.6 ml. The volume of substrate is kept constant at 2 ml and the volume of buffer adjusted so that the final mixture is 3 ml as previously. A progress curve for each volume of serum is prepared as previously described and the initial reaction velocity is determined. A graph is then plotted of enzyme activity and enzyme concentration expressed as millilitres of serum.

Choose a suitable time which is beyond the linear portion of the progress curve and express the activity as the average amount of p-nitrophenol liberated per minute. Plot this 'activity' against enzyme concentration and compare the result with the previous curve, when initial rates were used to express the activity.

Experiment 12.3   The effect of substrate concentration and inhibitors on ox heart lactate dehydrogenase

PRINCIPLE

Lactate dehydrogenase catalyses the reversible reduction of pyruvate to lactate with $NADH_2$ as the coenzyme. The reduced coenzyme ($NADH_2$) absorbs strongly at 340 nm while the oxidized form (NAD) does not, so the progress of the reaction can be followed by measuring the decrease in extinction at 340 nm with pyruvate as substrate. The theory of this method is discussed more fully in Chapter 7.

In this experiment, two compounds with similar structures to the natural substrate pyruvate are examined to see if they are competitive or non-competitive inhibitors.

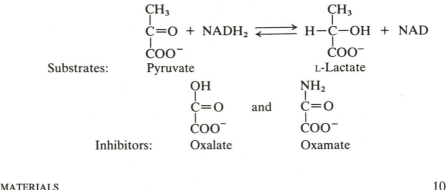

MATERIALS

		10
1. Phosphate buffer (0.1 mol/litre, pH 7.4)		1 litre
2. Sodium pyruvate (21 mmol/litre, prepared fresh in phosphate buffer)		200 ml
3. $NADH_2$ (3.5 mmol/litre, prepared fresh)		25 ml
4. Ox heart lactate dehydrogenase diluted in phosphate buffer when required		100 ml
5. Sodium oxalate (3 mmol/litre, in phosphate buffer)		50 ml
6. Sodium oxamate (15 mmol/litre, in phosphate buffer)		50 ml
7. Water bath at 37°C		5
8. Recording spectrophotometer		5

METHOD

Initial assay   Prepare the following mixture in duplicate and equilibrate at 37°C for 10 min.

Component	ml
Phosphate buffer	2.5
NADH$_2$ (3.5 mmol/litre)	0.1
Ox heart lactate dehydrogenase (diluted)	0.3

Rapidly add 0.1 ml of the sodium pyruvate solution also at 37°C, transfer to a cuvette in a thermostatically heated cell housing of an ultraviolet spectrophotometer and observe the change in extinction at 340 nm with time. Dilute the enzyme preparation so that an extinction change in the region of 0.05–0.10 per minute is obtained.

Calculation of the activity   Enzyme activities are expressed most conveniently in nanomoles per minute per millilitre of enzyme or nanomoles per minute per milligram of protein. The molar extinction coefficient of NADH$_2$ at 340 nm is 6.3×10^3 litres mol^{-1} cm^{-1} so that a solution of 1 μmol/ml has an absorption of 6.3.

For the 3 ml reaction mixture used in the assay:
Enzyme activity (μmol/min)
    = extinction change per min/6.3×3.
Since 0.3 ml of enzyme solution was used in the assay:
Enzyme activity (μmol min^{-1} ml^{-1})
    = extinction change per min/$6.3 \times 3 \times 1.0/0.3$
    = extinction change per min $\times 1.61$.
In general terms then enzyme activity (μmol min^{-1} ml^{-1})
    = E per min $\times 1000$/extinction coefficient NADH $\times$ volume in cuvette
      $\times 1.0$/volume used for assay.
The final activity is obtained by multiplying by any dilution factor for micromoles per minute per millilitre and dividing by the protein concentration in milligrams per millilitre for micromoles per minute per milligram.

Determination of the Michaelis constant   Dilute the sodium pyruvate solution ten times and repeat the above assay using volumes from 0.1 to 1.0 ml of the substrate to start the reaction. Adjust the initial volume of buffer in each case so as to give a final reaction mixture of 3 ml. Calculate the activity and plot graphs of v against s, $1/v$ against $1/s$ and s/v against s and determine the K_m and V.

The effect of inhibitors   (i) Repeat the above experiment but this time incorporate 0.1 ml of the inhibitor into the reaction mixture, again adjusting the volume of buffer to give a final reaction volume of 3 ml. Plot a graph of $1/v$ against $1/s$ and determine the inhibitor constant K_i. (ii) Repeat the initial assay but this time determine the activity at two fixed substrate concentrations and a range of concentrations of the inhibitor, up to a final concentration of 0.1 mmol/litre oxalate and 0.5 mmol/litre oxamate. If v_i is the activity in the presence of the inhibitor and v is the activity in the absence of the inhibitor, plot a graph of v/v_i against i. Use these graphs to decide what type of inhibition is given by oxalate and oxamate and determine the inhibitor constants.

High substrate concentration   Examine the effect of high substrate concentrations on a plot of v against s and s/v against s. This is carried out by using volumes up to 1 ml of 21 mmol/litre sodium pyruvate to trigger the enzyme reaction. Is there anything unexpected about the graphs? If so try to explain what is observed.

Experiment 12.4   The effect of temperature on the activity of α-amylase

PRINCIPLE

α-Amylase catalyses the hydrolysis of α 1–4 links of starch with the production of reducing sugars. The reaction is followed by measuring the increase in reducing sugars using the 3,5-dinitrosalicylate reagent when an alkaline solution of 3,5-dinitrosalicylic acid is reduced to 3-amino-5-nitrosalicylic acid. The reaction is followed by measuring the extinction at 540 nm.

This exercise and the next one are best carried out as class projects because of the large number of measurements to be taken. The quantities required are calculated on the assumption that each pair will assay in duplicate at one temperature only.

MATERIALS	100
1. Sodium or potassium phosphate buffer (0.1 mol/litre, pH 6.7)	5 litres
2. Buffered starch substrate (5 g/litre in phosphate buffer. Mix 5 g of soluble starch to a smooth paste with about 50 ml of buffer solution. Add this quantitatively to 500 ml of boiling phosphate buffer solution, continue to boil for 1 min, then cool to room temperature, and dilute to 1 litre with buffer solution)	3 litres
3. Sodium chloride (10 g/litre)	1 litre
4. α-Amylase suitably diluted	1 litre
5. Sodium hydroxide (2 mol/litre)	1 litre
6. Colorimeter	50
7. Water baths at a range of temperatures up to 75°C	30
8. Dinitrosalicylate reagent (Exp. 9.5)	1 litre

METHOD

(*a*) *Determination of the 'energy of activation'*   Set up eight tubes containing the following reaction mixture:

Contents	ml
Starch (5 g/litre)	2.5
Phosphate buffer (0.1 mol/litre, pH 6.7)	1.0
Sodium chloride (10 g/litre)	0.5

Place the tubes in a water bath at a fixed temperature and equilibrate the enzyme at the same temperature. After 10 min add 0.5 ml of the enzyme to seven of the tubes and 0.5 ml of water to the blank. Incubate the tubes for 0, 5, 10, 15, 20, 30, and 40 min and stop the reaction by adding 0.5 ml of 2 mol/litre NaOH to all tubes. Add 0.5 ml of the dinitrosalicylate reagent and heat the tubes for 5 min in a boiling water bath, cool, and read the extinction at 540 nm against a blank. (*Note:* The tubes must all be cooled to room temperature before reading since the extinction is sensitive to temperature change.)

Plot the extinction of each solution against the time of incubation and prepare a progress curve of the reaction. Collect the data from the other groups and plot a graph of the change in the initial reaction rate against temperature and also the substrate changed per minute at 10 and 40 min against temperature. Compare these graphs and explain them.

Plot a graph of $\log_{10}v$ against $1/T$ and determine the heat of activation. What assumptions have been made in this experiment?

For convenience the activity can be expressed in terms of extinction change at 540 nm per minute.

(*b*) *Temperature and enzyme stability*   Prepare a suitable dilution of the α-amylase so as to give a good activity when incubated at 37°C for 5 min. Place the enzyme in a boiling tube and equilibrate for 10 min. Pipette 0.5 ml of the α-amylase into four test tubes in a water bath maintained at a fixed temperature from 50 to 75°C. Remove the tubes after 5, 10, 15 and 20 min (or other suitable time intervals) and rapidly cool under running tap water. Incubate the tubes at 37°C and assay for enzyme activity after the addition of the substrate as in the previous experiment. The mixture is incubated for 5 min and the activity determined with 3,5-dinitrosalicylate. The rate of inactivation of the enzyme can then be determined. Collect the data from the groups who have looked at other temperatures and prepare a graph of the rate of inactivation (v) against $1/T$ and determine the heat of inactivation. Compare this figure with that obtained for the heat of activation from the last experiment.

Experiment 12.5   The determination of the molecular weight and purity of chymotrypsin from the enzyme kinetics

PRINCIPLE
The hydrolysis of certain carboxylic esters by chymotrypsin displays anomalous kinetics. Chymotrypsin catalyses the hydrolysis of *p*-nitrophenyl acetate and the

enzyme reacts initially with the substrate by becoming acetylated at a serine hydroxyl at the active site. This first stage is accompanied by an initial burst of *p*-nitrophenol after which the release of the product proceeds at a slow steady rate (i). The subsequent hydrolysis of the acetylchymotrypsin proceeds at a much slower rate (ii).

The amount of *p*-nitrophenol released in the initial burst is stoichiometrically related to the amount of enzyme present and can be used to calculate the minimum molecular weight of chymotrypsin.

(i) Release of *p*-nitrophenol

$$E-OH \ + \ CH_3COO-\hspace{-0.5em}\bigcirc\hspace{-0.5em}-NO_2 \ \longrightarrow \ E-OOC\cdot CH_3 \ + \ HO-\hspace{-0.5em}\bigcirc\hspace{-0.5em}-NO_2$$

(ii) Hydrolysis of acetylated enzyme

$$E-OOC\cdot CH_3 \ + \ H_2O \ \longrightarrow \ E-OH \ + \ CH_3COO^- \ + \ H^+$$

MATERIALS   <u>10</u>
1. Tris–H_2SO_4 buffer (0.1 mol/litre, pH 7.6)   500 ml
2. Chymotrypsin (0.8 mg/ml in the tris buffer)   50 ml
3. Stock *p*-nitrophenyl acetate (1 mmol/litre in ethanol, freshly prepared) 25 ml
4. Working *p*-nitrophenyl phosphate (0.1 mmol/litre in tris buffer, 100 ml
  prepared by diluting the stock solution 1 in 10 with the tris buffer)
5. *p*-Nitrophenol (50 μmol/litre in tris buffer)   100 ml
6. Recording spectrophotometers   5

METHOD

Prepare two cuvettes as below:

	Reference	Reaction
p-Nitrophenyl acetate (0.1 mmol/litre)	1.5 ml	1.5 ml
Tris–H_2SO_4 buffer (0.1 mol/litre, pH 7.6)	1.5 ml	—
Chymotrypsin (0.8 mg/ml)	—	1.5 ml
Water	0.5 ml	0.5 ml

Immediately after adding the enzyme, mix thoroughly and record the increase in extinction at 400 nm to obtain the linear part of the progress curve. Extrapolate the curve back to zero time and record the extinction.

Repeat the experiment with varying concentrations of chymotrypsin. Meanwhile prepare a standard curve of *p*-nitrophenol (0–50 μmol/litre) and use this to plot the μmoles of *p*-nitrophenol released at zero time against the concentration of chymotrypsin. From this calculate the minimum molecular weight of chymotrypsin.

Remember
1. Calculate the μmoles of *p*-nitrophenol in the cuvette.
2. Calculate the μg of chymotrypsin in the cuvette.
3. Mol. wt = mass/molarity expressed in the same units, i.e., mg and mmoles or μg and μmoles.

The true mol. wt of pure chymotrypsin is 24 500 but it is very unlikely that the value obtained by this method will be this and it will almost certainly come out larger than this. The actual value obtained depends on how good the sample of the enzyme is. Explain why this is so and use the data to determine the purity of the enzyme preparation.

Experiment 12.6   Yeast isocitrate dehydrogenase: an allosteric enzyme*

PRINCIPLE

Yeast isocitrate dehydrogenase catalyses the conversion of L-isocitrate to α-oxoglutarate and carbon dioxide with NAD as the coenzyme. It is a major control point in the tricarboxylic acid cycle and is modified by a number of *effectors* including NAD, AMP and Mg^{2+}. The enzyme is assayed by following the increase in extinction at 340 nm as NAD is reduced to $NADH_2$.

L-Isocitrate α-Oxoglutarate

MATERIALS	10
1. Fresh bakers yeast	150 g
2. Washed sand	—
3. Sodium bicarbonate (0.1 mol/litre)	200 ml
4. Tris–HCl buffer (30 mmol/litre, pH 7.4)	500 ml
5. Sodium D, L-isocitrate in tris buffer, pH 7.4	200 ml
6. NAD (2 mmol/litre) in tris buffer, pH 7.4	250 ml
7. AMP (3 mmol/litre) in tris buffer, pH 7.4	50 ml
8. $MgCl_2$ (0.1 mol/litre)	50 ml
9. Pestle and mortar	5
10. A recording ultraviolet spectrophotometer	5
11. Circulating water bath at 30°C	5

*By permission of Dr P. J. Butterworth.

METHOD

Preparation of the enzyme   Prepare a crude extract of the enzyme by grinding some freshly grown yeast with washed sand and sodium bicarbonate solution (0.1 mol/litre) in a pestle and mortar. The yeast cell wall is extremely tough and considerable grinding is required to break it. The proportions of yeast and sand are adjusted so that there is plenty of abrasive action by the sand, yet sufficient sodium bicarbonate present in which the released enzyme can dissolve. The grinding requires a little practice and it may be advisable for a technician to prepare a fresh extract for the class.

After grinding, transfer the mixture to centrifuge tubes, and remove the sand and large cell debris by centrifuging at 1000 g for 10 min. Collect the supernatant and recentrifuge at 30 000 g for 1 h. Discard the precipitate and store the supernatant on ice until required.

Enzyme assay   Prepare a series of tubes containing the following reaction mixture and incubate at 37°C for 10 min.

NAD (2 mmol/litre)	1.5 ml
MgCl$_2$ (0.1 mol/litre)	0.1 ml
Diluted enzyme	0.2 ml

The amount of enzyme required has to be determined by experiment and if the preparation is too active it should be diluted with tris buffer.

Start the reaction by adding 1.2 ml of substrate previously incubated at 37°C and follow the increase in extinction at 340 nm. Set up the appropriate blanks and controls to allow for any non-enzymic changes.

Experiment   Measure the enzyme activity over a range of substrate concentrations from 0.1 to 1.2 mmol/litre. This is achieved by triggering the reaction with from 0.1 to 1.2 ml of the substrate and adding tris buffer to give a final added volume of 1.2 ml. The substrate used is a mixture of the D and L forms so that the L-isocitrate concentration ranges from 0.05 to 0.6 mmol/litre.

Repeat the experiment but this time include AMP in the reaction mixture at final concentrations of 0.2 mmol/litre and 0.5 mmol/litre.

Express the enzyme activity in terms of nanomoles of NAD reduced per minute and prepare graphs of v against s, $1/v$ against $1/s$ and $\log v/(V-v)$ against $\log s$.

If time permits, repeat the above experiment after heating the enzyme at 60°C for 5 min. Another possible investigation is to examine the effect of Mg^{2+} concentration on the reaction.

Experiment 12.7   The isolation of muramidase (mucopeptide *N*-acetylmuramyl hydrolase, 3.2.1.17) from egg white

PRINCIPLE

The enzyme is a potent antibacterial agent and acts by catalysing the hydrolysis of β 1–4 links between *N*-acetylmuramic acid and *N*-acetylglucosamine residues present in the mucopeptide of the cell walls. This particular property forms the basis of the assay method when the enzyme is mixed with a turbid suspension of freeze-dried bacteria. As the hydrolysis proceeds, the turbidity of the suspension decreases and this can be readily followed in a spectrophotometer at 450 nm.

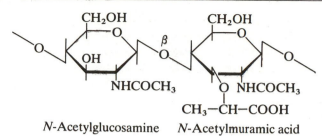

N-Acetylglucosamine   N-Acetylmuramic acid

Muramidase has a low molecular weight (14 300) and a high isoelectric point (pH 10.5) and this latter property is used to separate the enzyme from other unwanted proteins.

Muramidase was known as lysozyme for many years but this trivial name is no longer recommended.

MATERIALS $\dfrac{10}{15}$

1. Hens' eggs
2. Muslin   —
3. HCl (1 mol/litre and 0.1 mol/litre)   100 ml
4. Glass wool   —
5. Concentrated buffered saline (NaCl, 0.5 mol/litre, and tris–EDTA,   2 litres
   0.5 mol/litre, pH 8.2)
6. Cation exchange resin, CM-cellulose   60 g
7. Sodium carbonate–bicarbonate buffer (0.2 mol/litre, pH 10.5)   2 litres
8. NaCl   20 g
9. Acetic acid (1 mmol/litre)   10 ml
10. Standard albumin for protein assays (1 mg/ml)   100 ml
11. pH meter   5
12. Fraction collector   5
13. Ultraviolet and visible spectrophotometer   5
14. Water bath at 25 or 37°C   5
15. Sodium or potassium phosphate buffer (0.1 mol/litre, pH 7)   1 litre
16. Suspension of freeze-dried *Micrococcus lysodeikticus* in phosphate   500 ml
    buffer freshly prepared at 0.3 mg/ml)

METHOD

Enzyme and protein assay

Muramidase   Equilibrate the suspension of freeze-dried bacteria at 37°C and shake before using. Pipette 2.8 ml into a cuvette and start the reaction by adding 0.2 ml of the muramidase solution. Mix thoroughly and follow the decrease in absorbance at 450 nm in a recording spectrophotometer. Calculate the activity of the enzyme from the linear part of the progress curve and express the results in terms of micrograms of bacteria hydrolysed per minute.

Protein Prepare a standard curve of bovine serum albumin by measuring the absorbance at 280 nm of a range of solutions up to 1 mg/ml. The protein content at each stage is then determined by measuring the extinction of the solution at 280 nm and reading the protein value off the standard curve.

Purification

Preparation of chromatography column Equilibrate 10 g of CM-cellulose with the buffered saline, diluted 1 in 10, by gently stirring for 24 h; then prepare a chromatography column of this material as described in Chapter 4.

Purification table At each stage of the purification record the following values in the form of a table:

1. Volume (ml).
2. Enzyme activity (units/ml).
3. Total muramidase activity (units/ml × vol).
4. Total protein content (mg).
5. Specific activity (total muramidase activity/mg protein).
6. Purification factor for muramidase (specific activity divided by the initial specific activity).
7. Yield (per cent of original total activity).

Extraction Collect the whites from three eggs, filter through muslin to remove the chalazae, collect 50 ml of filtrate, and add this to 100 ml of water. Stir the mixture carefully without whipping air into the whites to prevent denaturation. Set aside 2 ml on ice for enzyme and protein assay.

pH precipitation Adjust the pH of the extract to 7.5 by slowly adding 1 mol/litre and 0.1 mol/litre HCl over a 10 min period. *Take care not to overshoot.* Remove the protein that precipitates by filtration through a glass wool plug in a filter funnel and add concentrated buffered saline to the filtrate to give a final concentration of 0.05 mol/litre NaCl and 0.05 mol/litre tris–EDTA, pH 8.2. Measure the pH and adjust to 8.2 if necessary. Record the volume and again set aside 2 ml on ice for the assay of muramidase and protein.

Column chromatography Carefully apply the extract to the column of CM-Sephadex and collect 10 ml fractions on a fraction collector. When the extract has run into the column, wash with a further 50 ml of dilute buffered saline before eluting with 0.2 mol/litre sodium carbonate–bicarbonate buffer. Continue collecting fractions until all the muramidase has come off the column.

Determine the protein and enzyme content of each fraction and plot their elution profiles. Calculate the per cent recoveries from the column.

Crystallization Muramidase is a strongly basic protein and readily forms crystalline salts with chloride, iodide, carbonate, etc. Under optimal conditions, crystallization proceeds slowly and reaches a maximum yield after 72–96 h.

Adjust the pH of the pooled enzyme peak to 10.5 and slowly add NaCl to a final concentration of 0.3 mol/litre. Leave to stand in the refrigerator for 3–4 days.

If crystallization has occurred, decant as much as possible of the supernatant and sediment the crystals from the remaining solution by centrifugation. Dissolve the precipitate in 1 ml of 1 mmol/litre acetic acid and remove any insoluble material by centrifugation. Determine the protein and muramidase content and record the final purification factor and yield.

Experiment 12.8   Separation of the isoenzymes of lactate dehydrogenase by electrophoresis on polyacrylamide-gel

PRINCIPLE

Definition   Enzymes which catalyse the same chemical reaction but which differ in certain of their physicochemical properties are known as *multiple molecular forms* and electrophoresis or chromatography is frequently used to separate and characterize the different forms.

If the multiple molecular forms are genetically determined, they are given the name of *isoenzymes*.

Origin of isoenzymes   Most animal tissues contain up to five isoenzymes of lactate dehydrogenase which can be readily separated by electrophoresis. They are numbered according to the speed of migration to the anode so that LD_1 is the fastest migrating species and LD_5 the form with the lowest mobility. The isoenzymes arise from various combinations of two subunits to form five possible tetramers. LD_1 contains only H subunits, so named as this is the predominant form in heart, while LD_5 contains only M subunits and is the main form present in skeletal muscle. The other three isoenzymes (LD_2, LD_3, LD_4) arise from the combination of these two subunits to form hybrid molecules (Fig. 12.10). The amount and distribution of these enzyme forms is characteristic of the tissue of origin and this is a useful way of fingerprinting the tissue.

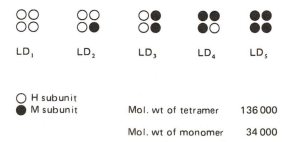

○ H subunit	Mol. wt of tetramer	136 000
● M subunit	Mol. wt of monomer	34 000

Fig. 12.10   The composition of lactate dehydrogenase isoenzymes

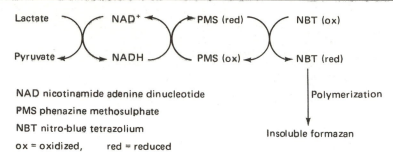

NAD nicotinamide adenine dinucleotide

PMS phenazine methosulphate

NBT nitro-blue tetrazolium

ox = oxidized,   red = reduced

Polymerization

Insoluble formazan

Fig. 12.11 Reactions involved in the detection of lactate dehydrogenase isoenzymes

Staining for enzyme activity   The isoenzymes of LDH are visualized by incubating the gels with a mixture containing the normal substrate and coenzyme, phenazine methosulphate (PMS) (an artificial electron acceptor), and a tetrazolium dye. On reduction, this dye forms a highly-coloured and sparingly-soluble formazan which precipitates at the sites of enzyme activity (Fig. 12.11).

MATERIALS

		10
1.	Solutions and equipment for polyacrylamide-gel electrophoresis as in Exp. 5.4	5
2.	Sodium or potassium phosphate buffer (0.05 mol/litre, pH 7.4)	1 litre
3.	Rat (200–300 g weight)	1
4.	Ox heart and rabbit muscle lactate dehydrogenase instead of rat tissues	—
5.	Tris–HCl buffer (0.1 mol/litre, pH 9.2)	1 litre
6.	Lithium lactate	10 g
7.	Nicotinamide adenine dinucleotide (NAD)	300 mg
8.	Nitro blue tetrazolium (NBT)	30 mg
9.	Phenazine methosulphate (PMS)	20 mg
10.	Developing reagent (0.1 mol/litre lithium lactate, 50 mg NAD and 5 mg NBT in 100 ml of tris–HCl buffer, pH 9.2. This solution is stable for a week in the dark at 4°C. Immediately before use add PMS (approx. 0.5 mg)	500 ml
11.	Incubator at 37°C	5

METHOD

Prepare eight rods of polyacrylamide gel as detailed in Exp. 5.4 and allow to set. Meanwhile, kill a rat and prepare homogenates of liver, kidney, skeletal muscle, heart and other tissues in 0.05 mol/litre phosphate buffer (0.5–1 g/10 ml). Remove the debris by centrifugation and store the supernatant on ice until required.

Add 50 μl of each sample in duplicate to the polyacrylamide rods and carry out electrophoresis as described earlier (Exp. 5.4). At the end of the run incubate the gels with the developing reagent at 37°C for up to 10 min in subdued light. Photograph the electrophorogram or make a careful scale drawing of the position of the bands. Carefully note the rate at which each band develops, the final intensity of the stain, and the degree of spacing between the isoenzymes.

If time permits, repeat the exercise with blood serum, haemolysed erythrocytes, small intestine and lung tissue.

Experiment 12.9   Some properties of lactate dehydrogenase isoenzymes

PRINCIPLE

The electrophoretic properties of isoenzymes reflect differences in their catalytic activity so that isoenzymes differ in such properties as their K_m, stability to heat and urea and the effect of inhibitors.

The following experiments can be carried out on the separated fractions from Exp. 5.5 but crystalline ox heart and rabbit muscle enzymes can be used to show the difference between two tissues of quite different isoenzyme content. Ox heart contains the fast-moving isoenzymes LD_1 and LD_2, while rabbit muscle contains largely LD_5.

MATERIALS   10
1. Materials for Exp. 5.5   —
2. Reagents for the assay of lactate dehydrogenase (Exp. 12.3)   —

METHOD

A few suggested experiments are outlined below which illustrate the difference in properties of the isoenzymes.

1. Determine the Michaelis constants with pyruvate and 2-oxobutyrate as substrates.
2. Examine the stability of the enzyme to heat by heating it for up to 30 min at 55–70°C.
3. Incubate the enzyme for 30 min with increasing concentrations of urea up to 6 mol/litre and compare the activity with the control. Plot a graph of v/v_i against urea concentration.
4. Determine the inhibitor constants with oxalate.

Further reading

Bergmeyer, H. U., *Principles of Enzymatic Analysis*. Verlag Chemie, New York, 1978.
International Union of Biochemistry, *Enzyme Nomenclature*. Academic Press, New York, 1984.
Moss, D. W., *Isoenzymes*. Chapman and Hall, London, 1982.
Palmer, T., *Understanding Enzymes*. Ellis Horwood, Chichester, 1981.
Price, N. C. and Stevens, L., *Fundamentals of Enzymology*. Oxford University Press, 1982.

13. Membranes

Composition of membranes

Membranes are clearly visible under the electron microscope and form an essential part of the cell structure. Simple *prokaryotic cells* such as *Micrococcus lysodeikticus* and other bacteria are seen to have a single membrane surrounding the cell. *Eukaryotic cells* present in higher organisms also have a cell membrane and, in addition, contain other membrane-bounded structures such as a nucleus, mitochondria and lysosomes. These membranes effectively separate cells from each other and also divide them into distinct aqueous regions. Membranes are also the basic structures incorporating many enzymes and transport systems. Most membranes consist of about 40 per cent lipid and 60 per cent protein together with some carbohydrate and ions.

Phospholipids

Most of the lipid present in membranes consists of phospholipid molecules associated together in a regular manner. This structural organization arises from the fact that phosoholipids have *hydrophobic* and *hydrophilic* regions in the same molecule; such compounds are said to be *amphipathic* (Fig. 13.1).

Most phospholipids are also *zwitterions* since the phosphate group carries a negative charge at neutral pH and the base a positive charge (choline phosphoglyceride, ethanolamine phosphoglyceride) or a positive and negative charge (serine phosphoglyceride). The other common polar group linked to the phosphate residue is inositol but this is uncharged. The structure of the polar head group thus determines the magnitude and distribution of the charge on the membrane.

The polar region of the phospholipid is hydrophilic and seeks an aqueous environment, while the hydrophobic part of the molecule tends to increase its entropy by expelling the water from its vicinity and associating together with hydrophobic regions of other molecules. Phospholipids thus orientate themselves on the surface of an aqueous solution so that the polar region lies in the water and the alkyl side chain in the air (Fig. 13.2(a)). A similar type of association occurs in membranes with the alkyl side chains associating to form a lipid bilayer (Fig. 13.2(b)).

Cholesterol and other lipids

Most of the other lipids found in membranes are also amphipathic molecules and may confer particular properties on the membrane. Many *glycolipids* appear to determine cell and tissue immunological specificity and to be responsible for the major part of the charge carried on the cell surface.

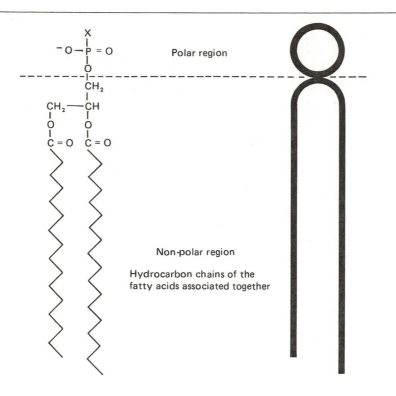

Fig. 13.1 The amphipathic nature of phospholipids

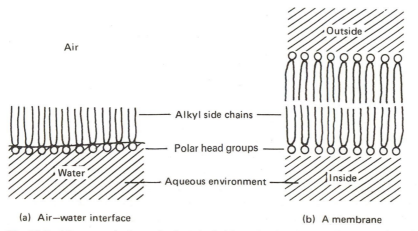

(a) Air–water interface

(b) A membrane

Fig. 13.2 The association of phospholipid molecules at an air–water interface and in a membrane

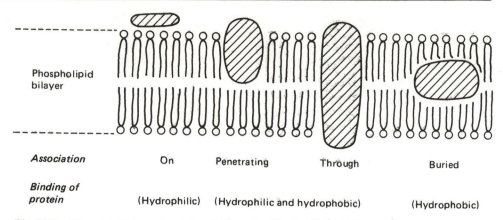

Fig. 13.3   The association of protein with the phospholipid bilayer in membranes

Proteins

Proteins are also present in membranes and are associated with the phospholipid bilayer in a variety of different ways (Fig. 13.3).

A few proteins, found on the membrane surface, are bound electrostatically to the bilayer and can be removed with high salt concentration. Most proteins, however, are more strongly bound to the membrane and, in many cases, can only be removed by destroying the membrane structure with detergents or organic solvents.

Membrane transport

Membranes carry out a variety of functions and some of these are highly specialized. Their principal role, however, is to control the transport of materials into and out of the cell and between the various cellular compartments.

Diffusion

Diffusion is a form of *passive transport* in which molecules move naturally from a higher to a lower concentration. The rate of transport is linear with concentration and does not show saturation (Fig. 13.4(a)).

Some examples of simple diffusion are the transport of ethanol, oxygen and carbon dioxide across the cell membrane.

Facilitated diffusion

This is also a form of passive transport down a concentration gradient but the rate of diffusion is greater than that expected from simple diffusion. Transport occurs by means of a *carrier* across the membrane and the process is also known as *carrier-mediated transport*. The rate of transport is hyperbolic with concentration and reaches a maximum value as the carrier becomes saturated (Fig. 13.4(b)).

Examples of facilitated diffusion include the passage of glucose and amino acids across membranes.

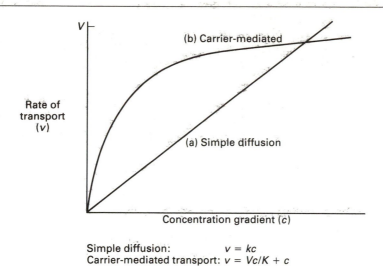

Simple diffusion: $v = kc$
Carrier-mediated transport: $v = Vc/K + c$

Fig. 13.4 The dependence of simple diffusion and carrier-mediated transport on the concentration gradient across the membrane

Active transport

In the case of active transport, compounds are transported against a concentration gradient with the expenditure of metabolic energy from the hydrolysis of ATP. Active transport is similar to facilitated diffusion in that it involves a carrier, shows a high degree of specificity and becomes saturated at high concentrations. The best known example of active transport is that of Na^+. The concentration of Na^+ inside the cell is much lower than the extracellular fluid and this concentration gradient is maintained by the Na^+ pump which removes Na^+ from the cell to the ECF using the energy obtained by the hydrolysis of ATP.

Experiments

Experiment 13.1   The effect of lipid composition on the permeability of a lipid monolayer

PRINCIPLE

The lipid composition of a membrane has a considerable effect on its permeability and, in this experiment, a lipid monolayer is used as a membrane model. If butanol is layered on top of water then two distinct phases are formed; if amphipathic lipids are present they will move into the boundary region. The polar part of the molecule will associate with the top aqueous layer and the hydrophobic region with the organic phase. Methylene blue is a highly coloured molecule and its passage across the boundary can be readily followed by eye. Unlike biological membranes, this model does not contain protein, but useful information can be obtained from this simple experiment as passive diffusion does depend on the lipid composition of the membrane.

MATERIALS   <u>100</u>
1. Fatty acids (stearic acid, oleic acid)   25 g
2. Acylglycerols (triolein, tripalmitin)   25 g
3. Phospholipid (egg lecithin)   25 g
4. Sterol (cholesterol)   25 g
5. Butanol   1 litre
6. Boiling tubes   700
7. Methylene blue in butanol (0.25 g/litre)   3 litres

METHOD

Set up seven boiling tubes each containing 5 ml of water. Carefully pipette 5 ml of butanol containing methylene blue and 200 mg of lipid down the side of each tube to form two distinct layers. Leave the tubes to stand at room temperature for 1–2 h and compare the results with that obtained using a control tube containing water but no lipid in the butanol. A rough estimate of the effectiveness of the lipids as permeability barriers can be obtained by measuring the extinction of the methylene blue in the aqueous phase.

Experiment 13.2   The effect of detergents and other membrane-active reagents on the erythrocyte membrane

PRINCIPLE

Many detergent-like molecules disrupt membranes by 'dissolving' the phospholipid components. In the case of erythrocytes, this effect can be readily followed in a colorimeter by measuring the absorbance of haemoglobin released from the disrupted cells.

MATERIALS   <u>10</u>
1. Fresh rat blood or time expired human blood used for transfusion   20 ml
2. Isotonic saline (8.9 g/litre)   2 litres
3. Detergents (10 g/litre)   5 ml
        Neutral, Triton X-100;
        Cationic, cetyltrimethylammonium bromide;
        Anionic, sodium dodecyl sulphate
4. Lysophosphatidyl choline (lysolecithin, 10 mmol/litre)   5 ml
5. Progesterone (100 mmol/litre in ethanol)   5 ml
6. Hydrocortisone (saturated solution in ethanol, approx 5 mmol/litre) 5 ml
7. Centrifuge tubes (10 ml)   100
8. Incubator at 37°C   3
9. Colorimeter or spectrophotometer   5

METHOD

Use human blood from a transfusion bottle or collect the blood from a freshly killed rat into a tube containing 0.5 ml of an anticoagulant (40 g/litre trisodium citrate).

Centrifuge the cells, wash them twice with isotonic saline and resuspend them in the same volume as the original blood. Dilute the erythrocyte suspension with saline so that when 0.5 ml is added to 4.5 ml of Triton X-100, an extinction of about 0.8–0.9 is obtained at 540 nm after centrifugation. This represents 100 per cent lysis and all subsequent extinction values should be expressed as per cent lysis.

Pipette duplicate 0.5 ml samples of the diluted erythrocytes into 10 ml centrifuge tubes containing 4.5 ml of saline. Mix thoroughly and add 50 μl of the test compound, mix by gentle swirling and place in an incubator at 37°C for 20 min. Separate any unbroken cells by centrifugation on a bench centrifuge and measure the absorbance of the supernatant solution at 540 nm.

First, examine a range of concentrations of the three detergents and prepare a graph of per cent haemolysis against concentration of the reagent. Repeat the experiment with lysolecithin, progesterone, and hydrocortisone and comment on the relative lytic effect of these compounds. Finally, examine the protective effect of hydrocortisone by incubating the erythrocytes with this steroid before exposing them to the lytic agent.

Experiment 13.3   The permeability of model membranes (liposomes) to anions

PRINCIPLE

Liposomes   Dry films of phospholipid swell spontaneously when in contact with aqueous solutions to form multilayered structures known as *liposomes*. These structures are made up of concentric bimolecular leaflets of phospholipid separated by small aqueous spaces (Fig. 13.5).

Sucrose is trapped within the liposomes as they form and, as this molecule cannot usually be passed out across the liposome membrane, it acts as an effective osmotic support. If the liposomes are suspended in an iso-osmotic solution of ions that cannot

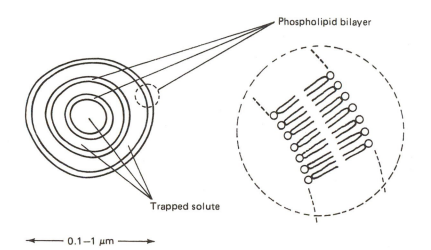

Fig. 13.5 Liposomes

penetrate the phospholipid bilayer then no change in the volume of the liposomes occurs. However, if the phospholipids are permeable to the ions, then swelling occurs as water moves into the liposomes with the ions to maintain osmotic equilibrium. The permeability of a solute can, therefore, be determined by following the light-scattering of a liposome suspension. When liposomes swell the absorbance decreases and when they contract the absorbance rises.

Membrane permeability   Liposomes are useful models for studying the permeability of the phospholipid part of membranes to anions. The aim of this experiment is to encourage the investigators to develop their own ideas of how anions may cross membranes from the results obtained. A few general points should be borne in mind when discussing the results.

1. Liposomes are virtually impermeable to cations (H^+, Na^+, K^+), unless ionophores are present.
2. Most species will cross membranes only in the un-ionized form.
3. If ion transport occurs, electrical neutrality must be maintained.

Ionophores   Many compounds modify ion transport and the antibiotic valinomycin is particularly interesting as it acts as a specific carrier of K^+. The molecule has a doughnut-like structure with the K^+ fitting neatly into the hydrophilic centre of the ring. The rest of the molecule is hydrophobic and, therefore, readily crosses the phospholipid of the membrane carrying the K^+ with it. Addition of valinomycin can, therefore, alter the course of liposome swelling in some cases. Molecules like this which facilitate the transport of specific ions across membranes are known as *ionophores*.

2,4-Dinitrophenol and other uncoupling agents act by discharging the proton gradient across membranes so that they become permeable to H^+.

MATERIALS			10
1. Choline phosphoglyceride			200 mg
2. Dicetyl phosphate			20 mg
3. Chloroform			50 ml
4. Sucrose (100 mmol/litre)			20 ml
5. Round-bottomed flask (100 ml)			5
6. Rotary evaporator			5
7. Test solutions (50 mmol/litre)			50 ml
NaCl	KCl	NH_4Cl	
NaI	KI	NH_4I	
NaCNS	KCNS	—	
CH_3COONa	CH_3COOK	CH_3COONH_4	
8. Valinomycin			100 μg
9. Nitrogen cylinders			2
10. 2,4-Dinitrophenol (5 mmol/litre in ethanol)			10 ml
11. Visible spectrophotometer with recorder			5

METHOD

Preparation of liposomes   Mix 20 μmol of choline phosphoglyceride with 2 μmol of dicetyl phosphate in about 3 ml of chloroform in a 100 ml round-bottomed flask. The dicetyl phosphate is present to give the liposomes a net negative charge. Remove the chloroform by rotary evaporation and add 3 ml of 100 mmol/litre sucrose to the dry lipid film. It is essential to remove all the chloroform before adding the sucrose and this is best checked by any odour in the flask. When the flask appears free of the solvent flush it out with nitrogen, add the sucrose and agitate the flask gently by hand to disperse the lipids.

Liposome swelling   The total absorbance change is small so the conditions for the swelling should first be optimized. This is done by adding 100 μl of the liposome suspension to 3 ml of distilled water and following the decrease in extinction at 450 nm in a recording spectrophotometer as the liposomes swell. Make any adjustments necessary to the chart speed, sensitivity, etc., then proceed with the investigation of anion permeability.

Add 100 μl of the liposome suspension to a cuvette containing 3 ml of one of the test solutions. Follow any change of absorbance with time. If no swelling occurs, make the liposomes specifically permeable to K^+ by adding 1 μg of valinomycin and continue to follow the absorbance at 450 nm. What effect does 2,4-dinitrophenol have on the swelling of liposomes?

In the light of your results, discuss the possible mechanisms by which the various anions may cross phospholipid membranes.

Experiment 13.4   The effect of cholesterol on the anion permeability of a phospholipid membrane

PRINCIPLE

Cholesterol modifies the packing of phospholipids and affects their mobility in the membrane. Cholesterol therefore changes the permeability properties of liposomes.

MATERIALS $\dfrac{10}{-}$
1. Reagents as in Exp. 13.3
2. Cholesterol   25 mg

METHOD

Prepare another batch of liposomes, but this time containing 10 μmol of cholesterol as well as 20 μmol of choline phosphoglyceride and 2 μmol of dicetyl phosphate. Select those salts which were permeable to the liposomes under the conditions of the experiment and repeat the experiment using the cholesterol-containing liposomes. Discuss your results and suggest how cholesterol affects the permeability of the liposomal membrane.

Experiment 13.5   The effect of insulin on the transport of glucose into isolated fat cells

PRINCIPLE

Glucose uptake   Insulin increases the rate of glucose transport across many cell membranes and this is illustrated in the following experiment by measuring the rate of removal of glucose from the incubation medium by isolated fat cells. In this case membrane transport is the rate-limiting step in glucose metabolism by the cells, so that an increase in the rate of removal of glucose from the medium is taken as indicating stimulation of glucose transport across the fat cell membrane.

Isolated fat cells   Fat cells are isolated from adipose tissue of the epididymal fat pads from male rats by digestion with collagenase. This enzyme catalyses the hydrolysis of the intercellular matrix and liberates isolated cells which, because of their high triglyceride content, float on gentle centrifugation. This procedure effectively isolates the fat cells from other cell types present in the tissue and a homogeneous population of cells can thus be obtained. These cells are metabolically active and respond to a wide range of physiological and pharmacological stimuli.

MATERIALS <u>10</u>

1. Sodium chloride (0.154 mol/litre) — 1 litre
2. Potassium chloride (0.154 mol/litre) — 50 ml
3. Calcium chloride (0.110 mol/litre) — 25 ml
4. Potassium phosphate (0.154 mol/litre) — 10 ml
5. Magnesium sulphate (0.154 mol/litre) — 10 ml
6. Sodium bicarbonate (0.154 mol/litre) — 250 ml
7. Krebs–Ringer bicarbonate with reduced calcium concentration (KRB). Mix solutions 1–6 in the following proportions then gas with 95 per cent O_2/5 per cent CO_2 for 20 min — 1285 ml
   - 1000 ml of solution 1
   - 40 ml of solution 2
   - 15 ml of solution 3
   - 10 ml of solution 4
   - 10 ml of solution 5
   - 210 ml of solution 6
8. KRB–albumin solution. Prepare a 30 per cent w/v solution of albumin in KRB and dialyse overnight against KRB in the cold. The following day, dilute to 4 per cent w/v albumin with KRB and gas to pH 7.4 with 95 per cent O_2/5 per cent CO_2. Store at 37°C in a sealed container under an atmosphere of 95 per cent O_2/5 per cent CO_2 — 500 ml
9. Incubation medium (KRB–albumin containing 3 mmol/litre glucose) — 100 ml
10. Gas cylinder of 95 per cent O_2/5 per cent CO_2 — 3
11. Male rats (200–300 g) — 10

12. Plastic centrifuge tubes (10–15 ml)   100
13. Collagenase   30 mg
14. Insulin stock solution. (Dissolve 8 mg in 1 ml of 3 mmol/litre HCl)   8 mg
15. Reagents for the assay of glucose using glucose oxidase (Exp. 9.6)   —
16. Shaking water bath at 37°C   5

Note   Plastic or siliconized glassware must be used throughout for all manipulations of the cells.

METHOD

Preparation of isolated fat cells   Kill two male rats and remove the four epididymal fat pads and place in 10 ml of KRB–albumin containing 5 mg of collagenase in a plastic tube. Gas briefly with 95 per cent O_2/5 per cent CO_2, seal the tube, and incubate by shaking for 1 h in a 37°C water bath. Remove any tissue fragments with forceps and centrifuge the cell suspension in a plastic centrifuge tube at 400 *g* for 1 min. Penetrate the layer of fat cells on the surface with a Pasteur pipette and remove the cells below (infranatant) and sedimented cells by aspiration. Resuspend the fat cells in 10 ml of KBR–albumin and wash by gentle stirring with a plastic rod. Centrifuge at 400 *g* for 1 min and again remove and discard the infranatant. Repeat the washing procedure twice more and finally resuspend the cells in a suitable volume of the incubation medium (KRB–albumin–glucose).

Glucose uptake   First, determine the dose–response curve for the stimulation of glucose uptake by insulin. This is carried out by incubating 1 ml of the fat cell suspension in triplicate with the following final concentrations of insulin: 0, 1, 10, 100, and 1000 μU/ml (1 mg = 25 units). The insulin is added in a small volume (5 μl) to the plastic incubation tubes followed by 1 ml aliquots of the stirred cell suspension using a wide-bore 1 ml plastic syringe. Gas with 95 per cent O_2/5 per cent CO_2 for 2 min and incubate in sealed plastic tubes for 2 h at 37°C with gentle shaking.

After incubation, centrifuge the cells and determine the glucose remaining in the incubation medium using the glucose assay (Exp. 9.6). Express the results as a mean number of micromoles of glucose taken up per millilitre of suspension per hour ± standard error of the three incubations. Plot a graph of glucose uptake against $\log_{10}$ insulin concentration to give the dose–response curve. What concentration of insulin gives half of the maximum stimulation of glucose uptake by the cells? How does this compare with the resting plasma insulin concentration of about 20 μU/ml?

Experiment 13.6   The transport of amino acids across the small intestine

PRINCIPLE

Transport of metabolites   After the digestion of food in the gut, the amino acids and other small molecules produced are absorbed from the intestine into the blood stream. In many cases the absorption is against a concentration gradient and thus involves active transport. The energy needed to drive such processes is provided by the generation of ATP, so that metabolic poisons that reduce or block ATP production will have an adverse effect on these transport processes.

The type of transport (active or passive) depends very much on the geometry of the molecule and this is seen in the absorption of the optical isomers of amino acids and sugars.

The everted sac   A convenient method for studying transport in the gut is to use a section of small intestine which is turned inside out and tied at each end. This *everted sac* is immersed in buffered ionic medium containing the metabolite, and changes in the concentration of the metabolite are measured after incubation. By turning the intestine inside out, transport is now from a large volume of the incubation medium into a small volume inside the everted gut. This, therefore, magnifies any absorption that occurs and is a more sensitive preparation than using the intestine in the normal way.

Histidine estimation   Histidine is assayed by reading the colour produced when the amino acid reacts with diazotized sulphanilic acid.

MATERIALS   10
1. Sodium chloride (0.154 mol/litre)   1 litre
2. Potassium chloride (0.154 mol/litre)   50 ml
3. Magnesium sulphate (0.154 mol/litre)   20 ml
4. Potassium phosphate buffer (0.1 mol/litre, pH 7.4)   250 ml
5. Krebs–Ringer–phosphate medium with calcium omitted (KRP).   1250 ml
   Mix the above solutions in the following proportions and gas out
   with 95 per cent O_2/5 per cent CO_2

Volume (ml)	Solution
1000	1
40	2
10	3
200	4

6. Incubation medium (KRP containing 18 mmol/litre glucose)   1250 ml
7. Gas cylinder (95 per cent O_2/5 per cent CO_2)   3
8. Rats   5
9. Glucose–saline (0.154 mol/litre NaCl containing 18 mmol/litre   1 litre
   glucose)
10. L-Histidine (5 mmol/litre in the KRP medium containing   500 ml
    18 mmol/litre glucose)

11. Inhibitor solutions. (Prepare in the incubation medium and gas
    with 95 per cent O_2/5 per cent CO_2. *Care:* Poison!)
    (a) Sodium cyanide (2 mmol/litre)   100 ml
    (b) Sodium iodoacetate (10 mmol/litre)   100 ml
    (c) Sodium iodoacetate (40 mmol/litre)   100 ml
    (d) Sodium malonate (5 mmol/litre)   100 ml
12. Acetic acid for deproteinization (0.35 mmol/litre)   250 ml
13. Sulphanilic acid (10 g/litre in 1 mol/litre HCl). If the blank is   100 ml
    coloured recrystallize the reagent
14. Sodium nitrite (50 g/litre, prepare fresh)   100 ml
15. Sodium carbonate (75 g/litre of the anhydrous salt)   100 ml
16. Ethanol (20 per cent v/v)   1 litre
17. Standard L-histidine (0.15 mmol/litre)   100 ml
18. Syringes (1 ml capacity with blunted needles)   10
19. Colorimeter   5
20. Shaking water bath at 37°C   5
21. Boiling water bath   5
22. Hardened filter paper, Whatman No. 54   —
23. Thread for ligatures   —
24. Glass rods with specially thickened ends (see Fig. 13.6)   5

METHOD

Preparation of the everted sac Kill the rat and open the abdomen by a midline incision. Remove the small intestine by cutting each end and manually stripping the mesentery. Wash the entire length of the small intestine with glucose–saline at room temperature to remove blood, debris, etc., and prepare the everted sac. Insert a narrow glass rod with thickening into one end of the intestine (Fig. 13.6(a)). Tie a ligature over the thickened part of the glass rod and evert the sac by gently pushing the rod through the whole length of the intestine (Fig. 13.6(b)). Carefully ligate both ends of the everted sac, remove the rod and place the intestine in a glucose–saline solution at room temperature. Tie off 2–3 cm lengths of intestine with thread and cut an open sac from the main length (Fig. 13.6(c)). Place a second ligature loosely round the open end of the sac and introduce a blunt needle attached to a 1 ml syringe. Tighten the loose ligature over the needle and inject 0.4 ml of the KRP–glucose solution for experiments measuring absorption rates or 0.4 ml of the amino acid solution for the concentration gradient experiments into the sac (Fig. 13.6(d)); tighten the ligature and withdraw the needle. All ligatures have to be firm enough to prevent leaks but not too tight so as to damage the tissue.

Amino acid transport Immerse the sac in 15 ml of the 5 mmol/litre histidine solution, gas briefly with 95 per cent O_2/5 per cent CO_2, seal the flask, and shake for 10 min in a 37°C water bath. At the end of this time, analyse 0.2 ml of the solution inside the everted sac (*serosal side*) and outside the sac (*mucosal side*) for L-histidine after deproteinization.

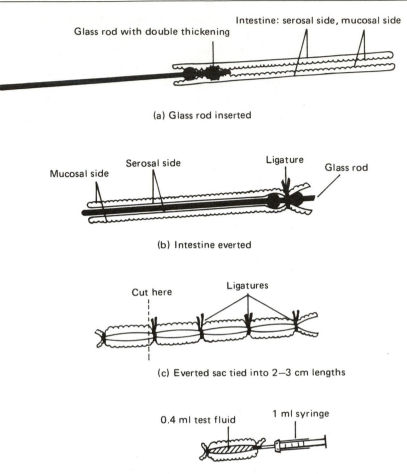

Glass rod with double thickening

Intestine: serosal side, mucosal side

(a) Glass rod inserted

Mucosal side   Serosal side   Ligature   Glass rod

(b) Intestine everted

Cut here   Ligatures

(c) Everted sac tied into 2—3 cm lengths

0.4 ml test fluid   1 ml syringe

(d) Segments of sac filled with test compound

Fig. 13.6 The preparation of everted sacs from rat small intestine

At the end of the experiment, empty the sac, blot with hardened filter paper (Whatman No. 54), and weigh the tissue after removing the ligatures.

Determination of histidine   Mix 2 ml of the weak acetic acid solution with 0.2 ml of the test solution in a test tube, cover with a marble, and place the tubes in a boiling water bath for 10 min. Cool and add distilled water to give a final volume of 5 ml. Filter or centrifuge as required.

Take 2 ml of the deproteinized solution, add 0.4 ml of sulphanilic acid (10 g/litre in 1 mol/litre HCl), mix thoroughly and add 0.4 ml of sodium nitrite (50 g/litre). Shake the tube and leave to stand for 5 min with occasional shaking. Add 1.2 ml of sodium carbonate solution (75 g/litre) and shake vigorously for about 10 s. Add 4 ml of ethanol (20 per cent v/v) and 2 ml of water, mix thoroughly and read the extinction at 498 nm against distilled water after 30 min. Calculate the concentration of histidine present by reference to a standard curve of histidine concentration using 0–2 ml of 0.15 mmol/litre histidine instead of the 2 ml of deproteinized fluid.

CALCULATION

1. Calculate the concentration gradient, which is the ratio of the amino acid concentration in the serosal fluid to that in the mucosal fluid.
2. Express the absorption rate as micromoles of amino acid per gram of wet tissue per hour ($\mu mol\, g^{-1}\, h^{-1}$).
3. Calculate the percentage recovery by comparing the amount found in the mucosal and serosal fluids with the amount taken originally.

FURTHER EXPERIMENTS

1. Examine the effect of various inhibitors upon the absorption of L-histidine.
2. Compare the absorption of the L and D forms of histidine.
3. Compare the absorption rates of various sugars using an assay for total carbohydrate (Exp. 9.4).

Experiment 13.7   The absorption of xylose from the gut in man

PRINCIPLE

Absorption of xylose   The following test is used in clinical chemistry to check for impaired absorption from the upper small intestine. The pentose sugar D-xylose is selected as this compound is not metabolized to any extent. Furthermore, there is no special mechanism for its reabsorption in the kidney, so the sugar is excreted in the urine. The test is carried out by determining the total xylose excreted in the 5 h following an oral dose of 5 g of the sugar. During this period, normal persons excrete 23–50 per cent of the dose, whereas those with steatorrhoea due to malabsorption excrete less than 20 per cent.

Xylose estimation   When xylose is heated in an acid solution, furfural is produced which reacts with *p*-bromoaniline to give a pink complex. The extinction of this complex is read in a colorimeter.

MATERIALS	100
1. *p*-Bromoaniline reagent. (Dissolve 20 g of *p*-bromoaniline in glacial acetic acid saturated with 4 g of thiourea then make up to 1 litre with acetic acid; shake well and filter)	6 litres
2. Pure D-xylose for human consumption	300 g
3. Standard D-xylose (0.1 g/litre prepared fresh)	1 litre
4. Water bath at 70°C	30
5. Colorimeter	50

METHOD

The test   Fast overnight, empty the bladder, and discard the urine, then drink a solution of 5 g of D-xylose dissolved in about 300 ml of water. Collect urine specimens every hour for 5 h and measure the volume and concentration of D-xylose.

Estimation of D-*xylose*   Dilute the urine 1 in 10 and 1 in 50, take 1 ml of each dilution and mix with 5 ml of the colour reagent in a test tube. Incubate for 10 min in a water bath at 70°C, cool to room temperature, and allow the colour to develop in the dark for 70 min. Repeat this using 1 ml of the standard D-xylose solution in place of the dilute urine, also set up a 'test control' and a 'standard control' by adding the 1 ml of diluted urine or 1 ml of standard to the *p*-bromoaniline solution and reading the colours immediately.

Carry out all the estimations in duplicate and read the extinctions at 524 nm.

CALCULATION

The concentration of the D-xylose solution is 0.1 mg/ml, therefore if V = volume of urine sample in millilitres and D the dilution, then the amount of xylose in the sample is given by:

$$(E_{\text{test}} - E_{\text{test control}})/(E_{\text{std}} - E_{\text{std control}}) \times 0.1 \times V \times D \text{ (mg)}$$

Plot histograms of the D-xylose excreted over each hour and calculate the percentage of xylose excreted over the test period.

Further reading

Carafoli, E. and Semenza, G., *Membrane Biochemistry — A Laboratory Manual on Transport and Bioenergetics.* Springer-Verlag, Berlin, 1979.

Harrison, R. and Lunt, G. G., *Biological Membranes*, 2nd edn. Blackie, Glasgow, 1980.

Houslay, M. D. and Stanley, K. K., *Dynamics of Biological Membranes.* Wiley, Chichester, 1982.

Finean, J. B., Coleman, R. and Michell, R. H., *Membranes and their Cellular Function*, 3rd edn. Blackwell Scientific Publications, Oxford, 1984.

West, I. C., *The Biochemistry of Membrane Transport.* Chapman and Hall, London and New York, 1983.

14. Cell fractionation

Cell structure

Basic cellular architecture

All living material is composed of cells and their products. The size and shape as well as the function of these cells vary widely so that, in one sense, there is no such thing as a typical cell although a great many cells have a number of features in common.

Under the light microscope, two distinct regions of the cell are visible, namely the nucleus and the cytoplasm which appears empty apart from a number of small particles. However, the electron microscope shows that the cytoplasm actually contains a number of quite distinct structures: Fig. 14.1 is a diagrammatic representation of some of the essential features of a 'typical animal cell' as seen with the electron microscope. It is not meant to be a drawing of an electron micrograph as this would be too complicated since the full extent of the endoplasmic reticulum and the total number of subcellular particles cannot be shown in a drawing of this type. However, the diagram does show certain basic features of cellular architecture common to a great many cells.

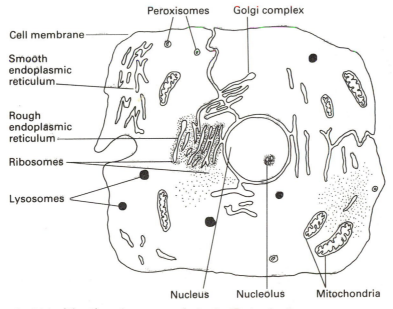

Fig. 14.1 The general structure of a 'typical' animal cell

Plasma membrane   Under the light microscope, the plasma membrane appears as a skin stretched over the cell with surface folds directed towards the interior (vesicles) and the exterior (microvilli). The overall thickness of the membrane is about 8 nm.

Endoplasmic reticulum   Many cells have a three-dimensional network of membranes known as the endoplasmic reticulum. Two types of membrane can be seen: the so-called *smooth endoplasmic reticulum* and *rough endoplasmic reticulum* named after their appearance under the electron microscope. The rough endoplasmic reticulum has small electron-dense particles called ribosomes attached to its surface, while the smooth form is free of ribosomes. The endoplasmic reticulum appears to be connected to the external nuclear membrane and the *Golgi complex*, the site of synthesis of a number of membranes.

Nucleus   Nearly all cells apart from bacteria contain a nucleus, a large structure of about 6 μm diameter which is clearly visible under the light microscope. Virtually all of the DNA of the cell is present in the nucleus complexed with histones in the form of nucleohistones. When the cell is not dividing, the DNA is distributed throughout the nucleus as chromatin, but during cell division the chromatin becomes organized into distinct linear structures.

Nucleolus   Within the nucleus there may be one or more distinct bodies of about 1 μm diameter which contain the bulk of the nuclear RNA.

Mitochondria   Mitochondria, which are the particles responsible for most of the oxidative metabolism in the cell, are just visible by light microscopy. The organelles from animal cells, plant cells and algae have a similar type of structure and under the electron microscope appear as cylindrical or spherical particles with a double membrane with the inner membrane invaginating into the cell to form *cristae*.

The size and number of mitochondria in a cell vary, but in rat liver there are about 800 of them occupying about 20 per cent of the cell volume. The size of mitochondria from different sources is very similar and those found in rat liver are from 1 to 2 μm long and have a diameter of 0.5 μm.

Lysosomes   Lysosomes are spherical particles about 0.5–1.0 μm diameter bounded by a single membrane. There are no obvious internal structures visible within the particles but they contain a whole range of hydrolytic enzymes which are involved in the digestion of exogenous and endogenous material.

Peroxisomes   Another group of membrane-bound particles about the same size as lysosomes contain the oxidative enzymes, catalase, urate oxidase and D-amino acid oxidase. These particles are known as *microbodies* or peroxisomes. The interior of these particles contains enzyme crystals which can occupy almost the whole of the internal volume.

Ribosomes   The ribosomes, which are the site of protein synthesis in the cell, occur free in the cytoplasm or attached to the endoplasmic reticulum. They are much smaller than the other organelles with a diameter of only 0.01–0.02 μm. Ribosomes from bacteria are 70 S particles while those from higher organisms are slightly larger with a sedimentation constant of 80 S. Ribosomes are made up of protein and RNA in about equal quantities and contain about 60 per cent of the total RNA in the cell. Each ribosome is made up of two oblate spheroids of unequal size which can be separated under low Mg^{2+} concentrations.

Practical cell fractionation

The metabolic function of the cell can be investigated to some extent by using histochemical means for detecting regions that contain a high activity of a particular enzyme. This approach has its limitations and it is more convenient to separate organelles and examine the properties of the isolated particles.

When cells are subjected to high shear, the cell membrane ruptures and the contents are released into the medium. By carefully controlling the conditions of homogenization, it is possible to avoid damaging the cell organelles, which can then be separated from each other by centrifugation. All steps must be carried out at 0°C to avoid damaging the particles.

Homogenization   A coaxial homogenizer consisting of a glass mortar and a hard Teflon pestle is very convenient.

The pestle is attached to an electric motor and small pieces of tissue suspended in the medium are placed in the glass mortar. The two parts of the homogenizer are brought together and the mortar is slowly moved up and down for about six to eight complete strokes while the pestle rotates at a controlled speed of about 2000 rev/min. As the homogenate is forced between the stationary wall of the mortar and the rotating pestle, the tissue is subjected to a shearing force which is sufficient to rupture the cells but not the organelles. These conditions are quite effective for liver but tougher tissues such as kidney and skeletal muscle may need first to be freed of connective tissue by forcing the tissue through a steel plate with holes under pressure before homogenization.

The clearance between the pestle and mortar, the speed of rotation of the pestle and the number of strokes all affect the preparation and must therefore be carefully defined, and thus one set of conditions which are suitable for a particular tissue cannot automatically be used for another tissue.

Suspending medium   The homogenizing medium should be cheap, uncharged and metabolically inert and for these reasons sucrose is the compound most frequently employed. For rat liver, a slightly hypotonic solution of sucrose (0.25 mol/litre) buffered with 20 mmol/litre tris to pH 7 has been found to be quite suitable. Ethylene diamine tetraacetic acid (EDTA) adjusted to pH 7 is sometimes incorporated into the medium at a concentration of 0.1 mmol/litre. This chelates calcium and other divalent ions which if present in even trace amounts can cause extensive swelling of mitochondria.

On the other hand, EDTA renders the mitochondrial membrane more permeable to monovalent ions so that some workers prefer to use sucrose alone.

Differential centrifugation   After homogenization, the suspension is separated into a number of fractions by centrifuging at various g values. The intracellular particles then sediment at different rates according to their mass.

The actual conditions of the fractionation depend on the tissue studied and those for the separation of rat liver mitochondria are not necessarily the same as those for the isolation of mitochondria from other rat tissue. Also, some fractions which are more or less homogeneous for one tissue may be very heterogeneous in others.

Density gradient centrifugation   Subcellular particles can also be separated by using differences in their density rather than mass. To do this, the homogenate is placed on top of a discontinuous gradient formed by layering a series of different sucrose concentrations on top of each other. The tubes are then centrifuged and, at equilibrium, the particles will be found as a band in that concentration of sucrose whose density is close to that of the organelles. This technique has been particularly useful in fractionating brain tissue when nerve endings and myelin can be isolated in a more or less homogeneous condition.

The alternative to a discontinuous density gradient is a continuous one, and relatively large quantities of material can be fractionated on such a gradient set up in a hollow centrifuge rotor, a technique known as *zonal centrifugation*.

Experiments

Experiment 14.1   The fractionation of rat liver

PRINCIPLE

Rat liver has probably been subjected to fractionation more times than any other material, and a more or less standard scheme for separating the subcellular particles is now available.

Centrifugation conditions		Major components in fraction
g value	time (min)	
500	5	Nuclei, whole cells, debris
8 000	10	Mitochondria, some lysosomes
15 000	10	Lysosomes, some mitochondria
100 000	60	Microsomes (membrane fragments, largely endoplasmic reticulum) and ribosomes
Final supernatant		Soluble components of the cell

MATERIALS   10
1. Isolation medium (0.25 mol/litre sucrose; 5 mmol/litre tris–HCl buffer,   2 litres
   pH 7.4; 0.1 mmol/litre EDTA)
2. Rats   5
3. Coaxial homogenizers   5
4. Ice baths   5
5. Ultracentrifuges   5

METHOD
Kill a rat, exsanguinate it and rapidly remove the liver. Wash the tissue free of blood in ice-cold sucrose, lightly blot and place in a tared beaker to weigh. Cut the liver into small fragments and homogenize in sucrose (20 g/100 ml) at 2000 rev/min by moving the mortar relative to the pestle for 8–10 complete strokes. Centrifuge the suspension in a refrigerated centrifuge according to the scheme shown in Fig. 14.2.

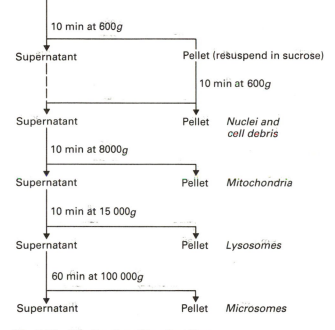

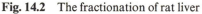

Fig. 14.2   The fractionation of rat liver

Ideally each fraction should be resuspended in sucrose and the washings combined with the supernatants. This has the advantage of producing purer fractions, but the disadvantage of introducing an increasing dilution of the cellular components.

Carefully resuspend the pellets in about 10 ml of sucrose and store on ice until required.

Experiment 14.2   The estimation of DNA, RNA and protein in the isolated cell fractions

PRINCIPLE

Nucleic acids   The tissue fraction is mixed with trichloracetic acid (TCA) to remove acid-soluble components, then extracted with ethanol to remove phospholipid. The lipid-depleted sediment is incubated overnight with warm alkali which hydrolyses the RNA to acid-soluble nucleotides but does not affect the DNA. On acidification, the DNA is dissolved in hot TCA before estimation. This method has the advantage that DNA and RNA are separated and can be estimated by determining the phosphorus,

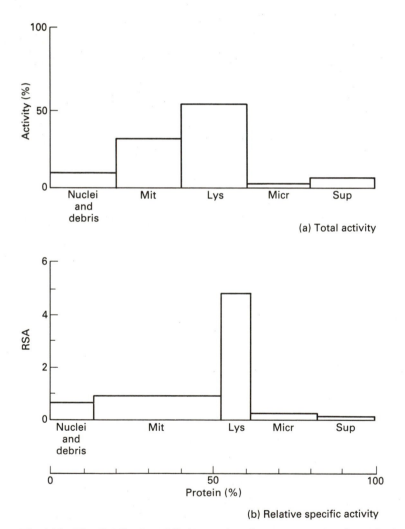

(a) Total activity

(b) Relative specific activity

Fig. 14.3   The distribution of β-glycerophosphatase among fractions obtained from rat liver by differential centrifugation

sugar or base content. The DNA is measured by the diphenylamine method and the RNA by the orcinol method as described earlier (Chapter 11).

Protein   Sucrose interferes with the biuret assay for protein and so the protein may have to be precipitated from the sample with TCA. If the sample needs to be diluted more than 1 in 50 for the assay then the TCA precipitation stage can be omitted. The detergent deoxycholate is incorporated into the biuret reagent to solubilize protein in association with lipid material so that the total protein content of each fraction is obtained.

Presentation of results   The total amount of DNA, RNA and protein in each fraction is calculated and the results presented as in Fig. 14.3. The specific concentration of the nucleotides is expressed as μg/mg protein.

MATERIALS   <u>10</u>
1. Trichloracetic acid (10 per cent w/v)   500 ml
2. Ethanol (95 per cent v/v)   1 litre
3. Potassium hydroxide (1 mol/litre)   250 ml
4. Hydrochloric acid (5 mol/litre)   50 ml
5. Reagents for the diphenylamine assay of DNA (Exp. 11.8)   —
6. Reagents for the orcinol assay of RNA (Exp. 11.9)   —
7. Biuret reagent with deoxycholate. (Dissolve 1.5 g of $CuSO_4 \cdot 5H_2O$ 1 litre
    and 6 g of Na, K tartrate in 100 ml of water. Add 300 ml of
    2.5 mol/litre NaOH and make up to 900 ml with water. Add 1 g of
    KI, 15 g of Na deoxycholate and make up to a final volume of 1 litre
8. Protein standard (bovine serum albumin 5 mg/ml)   50 ml
9. Boiling water baths   5

METHOD

Nucleic acids   Mix 2 ml of the fraction with 3 ml of 10 per cent TCA; centrifuge the precipitate and wash it with 3 ml of ice-cold TCA. Extract the precipitate three times with 5 ml of 95 per cent ethanol to give a lipid-depleted residue. Suspend the residue in 2 ml of 1 mol/litre KOH and incubate at 37°C overnight. The next morning, add 0.5 ml of 5 mol/litre HCl and 2 ml of 5 per cent TCA and centrifuge. The supernatant contains the acid-soluble nucleotides from RNA and the precipitate contains the DNA which is then dissolved in TCA by heating at 90°C for 10 min. Take suitable aliquots of the extracts and use them to estimate the nucleic acids as described in Chapter 11.

Protein   Take a suitable aliquot of the sample and add TCA to a final concentration of 5 per cent. Centrifuge in a bench centrifuge and decant off the supernatant. Add 1 ml of distilled water to resuspend the precipitate and add 2 ml of the biuret reagent; mix and heat in a boiling water bath for 1 min. Cool and read the extinction at 540 nm against a blank prepared by adding the biuret reagent to 1 ml of distilled water. Compare the extinction obtained with a series of protein standards of BSA (0–5 mg/ml) and construct a standard curve.

If the sample remains cloudy after incubation, shake the solution with 2 ml of diethyl ether and read the extinction of the aqueous phase.

Experiment 14.3   The distribution of enzymes in the cell

PRINCIPLE

Enzyme markers   Some enzymes are located almost exclusively in one part of the cell and this fact is used to check the 'purity' of a particular preparation of cell organelles. Such enzymes are known as marker enzymes and some of those that can be used for this purpose are shown below:

Fraction	*Marker enzyme*
Mitochondria	Glutamate dehydrogenase
Lysosomes	Acid phosphatase
Microsomes	Glucose-6-phosphatase
Supernatant	Lactate dehydrogenase

Errors can arise where the fractions are cross-contaminated and also where organelles become damaged by careless handling. This particularly applies to the acid phosphatase of lysosomes, which is released into the supernatant if the particles are damaged by adverse physical or chemical conditions.

Falsely low levels may be obtained for total enzyme activity unless the organelle is first rendered permeable to the substrate or even ruptured so that the enzyme is released into the medium.

Temperature of enzyme assays   All the enzymes should be assayed at 37°C but this requires UV spectrophotometers with a cell housing that can be maintained at this temperature. Such equipment may not be available in sufficient quantities for class work so the methods for the enzyme assays suggest room temperature for convenience.

Storage of fractions   It is best to carry out all the estimations on the same day but if this is not possible, then aliquots of each fraction can be stored in the deep freeze and the dehydrogenases measured later. However, glucose-6-phosphatase should be measured as soon as possible after the preparation of the fractions as this enzyme is unstable. It will probably also be convenient to measure the acid phosphatase on the day of fractionation.

METHOD

Measure the activity of all four enzymes in each of the fractions including the homogenate and also measure the protein content of each fraction as described in the last experiment. All enzyme assays must be done in duplicate and should be repeated if the two results differ markedly. It is suggested that assays are carried out on a 1 in 10 and 1 in 100 dilution of the fractions.

From these data calculate the total enzyme activity in each of the fractions and illustrate your results by expressing the activity in each fraction as a percentage of the total (Fig. 14.3(a)). In order to do this you will need to know the dilution of the sample, the volume of this used in the assay and the total volume of each fraction. Remember to allow for the fact that a sample of the homogenate is removed at the beginning when you come to calculate the recovery of each enzyme in the subcellular fractions.

Also express your results by plotting the *relative specific activity* (RSA) of the enzyme in each fraction against the protein in each fraction expressed as a percentage of the total (Fig. 14.3(b)). The RSA is obtained by dividing the specific activity in the fraction by that in the homogenate and is a useful indicator of the location of the enzyme and its 'purity' in the fraction. This second plot (Fig. 14.3(b)) also shows the enzyme in that fraction as a percentage of the total from the area of the rectangle.

Given that the marker enzymes are located exclusively in the fractions, comment on the efficiency of the fractionation procedure used in this experiment.

Experiment 14.3a   Glucose-6-phosphatase

PRINCIPLE

Glucose-6-phosphatase catalyses the hydrolysis of glucose-6-phosphate to glucose and phosphate. The phosphate is then estimated by the method previously described (Exp. 7.3). EDTA is incorporated into the reaction mixture to chelate Mg^{2+} required for alkaline phosphatase activity.

MATERIALS   <u>10</u>
1. Sodium cacodylate buffer (0.1 mol/litre, pH 6.5). *Care*: poison!   100 ml
2. Ethylene diamine tetraacetic acid (10 mmol/litre in buffer adjusted to   25 ml
   pH 6.5)
3. Glucose-6-phosphate (50 mmol/litre in buffer)   50 ml
4. Trichloracetic acid (10 per cent w/v)   250 ml
5. Reagents for phosphate estimation (Exp. 7.3)   —

METHOD

Prepare the following mixture and equilibrate at room temperature for 10 min:

Component	ml
Cacodylate buffer	1.2
EDTA	0.2
Glucose-6-phosphate	0.4

Add 0.2 ml of suitably diluted tissue fraction to start the reaction and incubate for 0, 10

and 20 min. Stop the reaction by adding 1 ml of ice-cold TCA (10 per cent). Centrifuge the precipitate and remove 1 ml of the supernatant for phosphate estimation.

Calculation   Check that the reaction is linear and calculate the activity as follows. The phosphorus content of 0.322 μmol of glucose-6-phosphate is 10 μg, so that 1 μg of phosphorus arises from 0.0322 μmol of glucose-6-phosphate.

In each test, 0.2 ml of diluted tissue fraction was taken and made up to 3 ml. From this solution, 1 ml was taken for phosphate estimation so that the final dilution of the tissue is: initial dilution $\times$ 3/0.2.

Set up suitable blanks and express the final results as micromoles of glucose-6-phosphate hydrolysed per minute.

Experiment 14.3b   Glutamate dehydrogenase

PRINCIPLE

Glutamate dehydrogenase catalyses the reversible oxidative deamination of glutamate to 2-oxoglutarate. NAD is required as coenzyme so that the reaction can be readily measured by following the change in extinction at 340 nm. EDTA is present in the reaction mixture to remove heavy metal ions which might otherwise inactivate the enzyme.

Maximum activity of the enzyme is only obtained after repeated freezing and thawing of the mitochondria or following the addition of detergent, hence the addition of Triton X-100.

MATERIALS   <u>10</u>
1. Sodium phosphate buffer (0.1 mol/litre, pH 7.4)   250 ml
2. Sodium-2-oxoglutarate (0.15 mol/litre prepared in the buffer and   25 ml
    adjusted to pH 7.4)
3. Ammonium acetate (0.75 mol/litre in phosphate buffer adjusted to   25 ml
    pH 7.4)
4. EDTA (30 mmol/litre in buffer adjusted to pH 7.4)   25 ml
5. NADH (2.5 mg/ml in phosphate buffer prepared fresh)   10 ml
6. Ultraviolet spectrophotometers   5
7. Triton X-100 (10 w/v in buffer)   25 ml

METHOD

Prepare the following mixture and equilibrate at room temperature for 10 min.

Component	ml
Phosphate buffer	2.1
Tissue fraction	0.2
NADH	0.1
Ammonium acetate	0.2
EDTA	0.2
Triton X-100	0.1

Start the reaction by adding 0.1 ml of the substrate 2-oxoglutarate, and follow the rate of change of extinction at 340 nm with time.

Calculate the enzyme activity as micromoles or nanomoles NADH oxidized per minute.

Molar extinction coefficient NADH $= 6.3 \times 10^3$ litres mol^{-1} cm^{-1}.

Experiment 14.3c   Acid phosphatase

PRINCIPLE

The principle is the same as for the assay of alkaline phosphatase (Exp. 12.1) with the difference that the alkaline phosphatase activity can be followed continuously whereas the acid phosphatase can only be determined after a fixed incubation time. This is because the product of the reaction, p-nitrophenol, does not absorb at 405 nm in the acid incubation buffer and is only coloured in alkaline medium. The reaction is therefore stopped by the addition of an alkaline buffer containing 0.4 mol/litre phosphate. The high phosphate concentration effectively inhibits any alkaline phosphatase activity that may be present giving a stable colour.

The linearity of the enzyme reaction must be checked. Enzyme assays should never rely on only a single timed incubation with the assumption that linearity occurs. This can be checked by incubating one sample for 10 min and another for 20 min. If the response is linear then the extinction reading for 20 min should be twice that found at 10 min. If this is not the case, then the experiment should be repeated with a different enzyme dilution until a linear response is obtained.

MATERIALS	10
1. Acetate buffer (0.2 mol/litre, pH 4.5)	250 ml
2. p-Nitrophenyl phosphate (8 mmol/litre, prepared fresh on the day of use)	100 ml
3. Tris–HCl buffer (1 mol/litre, pH 9.0 containing 1 mol/litre Na$_2$CO$_3$ and 0.4 mol/litre K$_2$HPO$_4$)	250 ml
4. p-Nitrophenol standard (50 µmol/litre; prepare a 5 mmol/litre stock solution in the alkaline tris buffer then dilute this solution 1 in 100 with the buffer)	100 ml
5. Triton X-100 (10 per cent w/v)	25 ml

METHOD

Add 0.2 ml of the enzyme solution to 1.2 ml of the acetate buffer and 0.1 ml of Triton X-100 and add 0.5 ml of the substrate solution with thorough mixing. Incubate for 10 min and 20 min and stop the reaction by adding 2 ml of the alkaline tris buffer.

Read the extinction at 405 nm and calculate the enzyme activity by reference to a standard curve of p-nitrophenol. If the extinction at 20 min is not double that at 10 min then the progress curve is non-linear and the assay must be repeated with a lower dilution of the tissue extract.

Prepare the appropriate blanks and allow for them in the calculation of enzyme activity.

Experiment 14.3d   Lactate dehydrogenase
The details for assaying this enzyme are given in Exp. 12.3.

Experiment 14.4   Mitochondrial swelling

PRINCIPLE

Changes in the volume of mitochondria suspended in isotonic solution can be brought about by a wide variety of agents including calcium ions, phosphate, arsenate, thyroxine and the higher fatty acids. The swelling action of these compounds is complex and depends on a number of factors such as the presence and absence of substrate, ADP, ATP, uncouplers and inhibitors. Some compounds act by stimulating the production of fatty acids from the mitochondrial membrane and bovine serum albumin which binds fatty acids therefore blocks their swelling action.

A large number of compounds can cause mitochondrial swelling but only ATP in the presence of Mg^{2+} can induce contraction of mitochondria and extrusion of water.

Mitochondrial swelling is shown by a fall in extinction and contraction by a rise.

MATERIALS	10
1. Materials for the preparation of rat liver mitochondria (Exp. 14.1)	—
2. Potassium chloride–tris solution (0.125 mol/litre KCl–0.02 mol/litre tris–HCl, pH 7.4)	500 ml
3. Bovine serum albumin (5 g/100 ml in KCl–tris)	10 ml
4. 2,4-Dinitrophenol (1 mmol/litre in KCl–tris)	10 ml
5. Thyroxin (1 mmol/litre in KCl–tris)	10 ml
6. ATP–$MgCl_2$ (0.06 mol/litre ATP, 0.15 mol/litre $MgCl_2$ in KCl–tris)	25 ml
7. Calcium chloride (15 mmol/litre in KCl–tris)	10 ml
8. Sodium phosphate (0.5 mol/litre adjusted to pH 7.4)	10 ml
9. Visible spectrophotometers	—

METHOD

Fractionate rat liver as in Exp. 14.1 and wash the mitochondrial pellet twice with the isolation medium by resuspension and centrifugation at 8000 g for 10 min. Suspend the final pellet in about 5 ml of sucrose and store on ice until required.

Immediately before use, dilute the suspension so that 0.1 ml added to the KCl–tris solution gives an initial extinction in the range 0.4–0.7. When a suitable dilution is found, add 0.1 ml to the mixtures below and follow the extinction at 520 nm with time. Take readings every half minute for the first 3 min, then continue to read at convenient time intervals for about 10 min. At the end of this time add 0.1 of the ATP–$MgCl_2$ mixture and continue to read for a further 5 min.

Component	Experiment no.					
	1	2	3	4	5	6
KCl–tris	2.9	2.8	2.7	2.8	2.8	2.8
Thyroxin	—	0.1	0.1	—	—	—
Bovine serum albumin	—	—	0.1	—	—	0.1
Sodium phosphate	—	—	—	0.1	—	—
Calcium chloride	—	—	—	—	0.1	—
Mitochondria	0.1	0.1	0.1	0.1	0.1	0.1
ATP–MgCl$_2$ added after 10–15 min	0.1	0.1	0.1	0.1	0.1	0.1

Finally investigate the swelling effect of a range of concentrations of the uncoupler 2,4-dinitrophenol. What effect does this compound have on the swelling action of other reagents? Compare the effect of adding the DNP before or after the swelling agent.

Experiment 14.5   The determination of lysosomal integrity

PRINCIPLE

The latency of lysosomal enzymes   Lysosomes are bounded by a single lipoprotein membrane which is normally impermeable to the enzymes and their substrates. If some of the lysosomes are broken by careless handling or other causes, then the enzymes are released into the medium and can be assayed. The degree of damage can be determined by measuring this *soluble or free activity* after removing the intact lysosomes by centrifugation and comparing this with the *total activity* measured after deliberately breaking the membrane with detergent. The difference between these values is known as the *latent activity* and this is usually expressed as a percentage of the total.

$$\text{Per cent latency} = (\text{total} - \text{free})/\text{total} \times 100$$

Under certain conditions, the membrane becomes permeable to the substrate but not to the enzyme so that we can speak of the *available activity*.

The meaning of these terms is shown diagramatically in Fig. 14.4.

Rat kidney lysosomes   Kidney lysosomes have been chosen for this and the following experiment since they can be obtained in a more pure state than lysosomes prepared from rat liver. On high speed centrifugation, two distinct layers can be seen in the pellet. The upper light-brown layer of a loose consistency is the mitochondrial fraction and the lower layer of a small dark-brown button of a sticky consistency are the lysosomes. The mitochondria can be removed by careful washing and the lysosomes

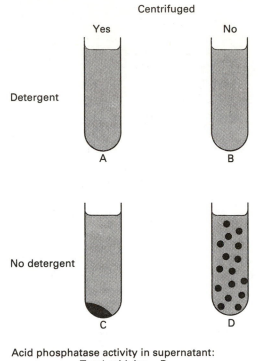

Acid phosphatase activity in supernatant:
          Total acitivity = B
          Free activity = C
          Available activity = D − C
          Latency % = (B − C)/B × 100

Fig. 14.4   The effect of Triton X-100 on the lysosomal membrane

obtained in a reasonable state of purity. Another advantage of kidney lysosomes is that they are tougher than liver lysosomes and can take more rough treatment which is always an advantage for a class experiment. However, they are still fragile structures and should be handled carefully and left on ice until required for the experiment.

MATERIALS	<u>10</u>
1. Isolation medium (0.35 mol/litre sucrose buffered to pH 7 with 10 mmol/litre triethanolamine butyrate)	500 ml
2. Rats (about 200 g body weight)	5
3. Reagents for the assay of acid phosphatase (Exp. 14.3c)	—
4. Triton X-100 (2 g/100 ml)	100 ml
5. Ice baths	5
6. Refrigerated ultracentrifuges	5
7. Recording visible spectrophotometers	5
8. Coaxial homogenizers	5

METHOD

Preparation of lysosomes Kill two rats and remove their kidneys. Decapsulate them and prepare an 8–10 per cent w/v homogenate in the isolation medium using six up and down strokes of the homogenizer rotating at 1500 rev/min.

Centrifuge the homogenate at 500 g for 5 min to sediment nuclei, intact cells and debris and resuspend the sediment. Centrifuge the suspension at 500 g for 5 min, combine the supernatants and centrifuge at 10 000 g for 5 min.

Remove the supernatant and discard. Add a small volume of the sucrose solution and gently swirl with a Pasteur pipette until the upper mitochondrial layer is suspended but the lower lysosomal pellet remains intact. Remove the mitochondrial suspension and resuspend the lysosomal pellet in the isolation medium. Centrifuge at 10 000 g for 5 min and repeat the washing procedure. Finally, resuspend the lysosomal pellet by gentle homogenization by hand in the buffered sucrose solution.

Determination of lysosomal integrity Take 0.5 ml aliquots of the lysosomal preparation and place into six small volume high speed centrifuge tubes. Add 0.5 ml of Triton X-100 or the isolation medium as indicated and centrifuge tubes A, C and E at 10 000 g for 10 min.

Tube	Lysosomes (ml)	2 per cent Triton X-100 (ml)	Isolation medium (ml)	H_2O (ml)	Centrifuged
A	0.5	0.5	—	—	+
B	0.5	0.5	—	—	—
C	0.5	—	0.5	—	+
D	0.5	—	0.5	—	—
E	0.5	—	—	0.5	+
F	0.5	—	—	0.5	—

Assay the acid phosphatase activity of either the suspension or the supernatant and determine the latency of the preparation.

What effect does water have on the lysosomes compared to detergent? If the activity in A is different from that in B explain the results.

Experiment 14.6 The effect of detergents on the stability of the lysosomal membrane

PRINCIPLE

Light scattering The physics involved in the scattering of light by biological particles is not simple, particularly when the particles are about the same size as the wavelength of the light. However, light scattering measurements can be of use in following rapid changes in lysosomes which cannot be conveniently determined by other means. In this investigation, measurements are taken at 520 nm and a decrease in the extinction of a suspension of lysosomes is taken to mean swelling or rupture of the particles. For light scattering measurements to be valid, the lysosomal preparation must contain a minimum of contaminating material and for kidney lysosomes this is reasonably true. However, changes in the extinction at 520 nm should be correlated with lysosomal enzyme release whenever possible.

Acid phosphatase   The effect of detergents on the lysosomal membrane is also investigated by measuring the release of the enzyme acid phosphatase from the organelles.

MATERIALS   <u>10</u>
1. Materials for the preparation of rat kidney lysosomes (Exp. 14.5)   —
2. Detergent solutions prepared in the buffered sucrose solution   100 ml
   (20 mmol/litre, except Triton X-100 which is 2 per cent w/v)

Detergent	*Charge*	
Cetyltrimethylammonium bromide	Cationic	—
$CH_3(CH_2)_{15}(CH_3)_3N^+$ Br^-		
Sodium dodecyl sulphate	Anionic	+
$CH_3(CH_2)_{11}OSO_3^-$ Na^+		
Lysophosphatidyl choline	Zwitterionic	+ —

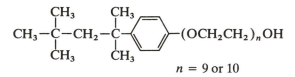

Triton X-100	Non-ionic	0

$$CH_3-\underset{\underset{CH_3}{|}}{\overset{\overset{CH_3}{|}}{C}}-CH_2-\underset{\underset{CH_3}{|}}{\overset{\overset{CH_3}{|}}{C}}\text{———}(OCH_2CH_2)_nOH$$

$$n = 9 \text{ or } 10$$

METHOD

Light scattering   Adjust the dilution of part of the lysosome preparation so that the addition of 0.1 ml of the suspension to 2.0 ml of buffered sucrose solution gives an initial absorbance in the range 0.3–0.5 at 520 nm when read against a sucrose blank.
     Examine the effect of a range of concentrations of the detergents on the light

scattering properties of the lysosomal suspension. Initially examine concentrations of detergent over a wide range then select a narrower range where an effect is observed. Suggested concentrations to start with are 5, 0.5, 0.05 and 0.005 mmol/litre. What happens if the buffered sucrose is replaced with distilled water?

Acid phosphatase   From your data select appropriate concentrations of the detergents to see what effect they have on the release of acid phosphatase on the lysosomes. An aliquot of the lysosomal suspension (0.5 ml) is mixed with an equal volume of the detergent and the lysosomes removed by centrifugation. The supernatant is then removed and assayed for acid phosphatase activity as in Exp. 14.5.

Experiment 14.7   The fractionation of pig brain by density gradient centrifugation

PRINCIPLE

Heterogeneity of brain tissue   The cellular structure of the brain is very heterogeneous so that homogenization and differential centrifugation give rise to subcellular fractions of a different composition to those from rat liver (Fig. 14.5). Each of these fractions contains a highly complex mixture of components and density gradient

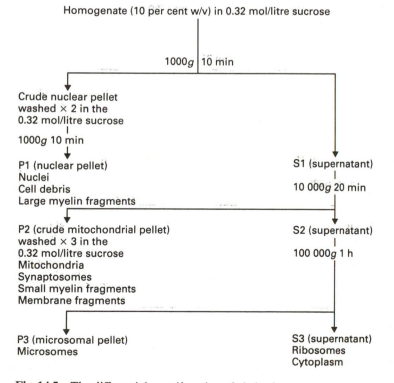

Fig. 14.5   The differential centrifugation of pig brain

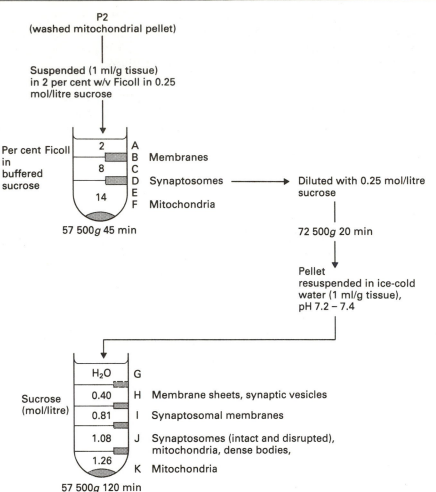

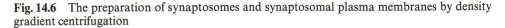

Fig. 14.6   The preparation of synaptosomes and synaptosomal plasma membranes by density gradient centrifugation

centrifugation is required to purify them further. This is illustrated in the following experiment when the mitochondrial pellet obtained by differential centrifugation is resolved into its constituents by centrifugation on a discontinuous gradient of Ficoll (Fig. 14.6).

Synaptosomes   During the initial homogenization, the pre-synaptic nerve endings shear off and the membranes reseal to form distinct particles known as synaptosomes (Fig. 14.7). These nerve-ending particles can be purified by density gradient centrifugation and their contents separated on a sucrose gradient following their disruption by osmotic shock (Fig. 14.6).

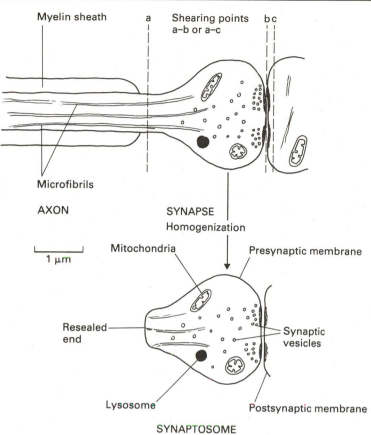

Myelin sheath   a   Shearing points
                       a–b or a–c   b c

Microfibrils

AXON

SYNAPSE
Homogenization

Mitochondria   Presynaptic membrane

1 μm

Resealed
end   Synaptic
                      vesicles

Lysosome   Postsynaptic membrane

SYNAPTOSOME

Fig. 14.7   The formation of synaptosomes during the homogenization of brain

MATERIALS   10
1. Pig brains obtained fresh from the slaughterhouse   10
2. Isolation medium (0.32 mol/litre sucrose, 5 mmol/litre tris–HCl, pH   1 litre
   7.4)
3. Buffered sucrose (0.25 mol/litre sucrose, 5 mmol/litre tris–HCl)   1 litre
4. Ficoll solutions (2, 8 and 14 per cent w/v in the buffered sucrose   100 ml
   solution)
5. Sucrose solutions for density gradients (0.4 mol/litre, 0.81 mol/litre,   50 ml
   1.08 mol/litre and 1.26 mol/litre adjusted to pH 7.2 – 7.4 with
   0.1 mol/litre NaOH)
6. Reagents for electron microscopy (details from the EM technician)   —
7. Coaxial homogenizers (clearance 0.25 mm)   5
8. Ultracentrifuges   5
9. Ice baths   5

METHOD

Fractionation of cerebral cortex   Strip the brain of its external membranes, wash with ice-cold isolation medium and remove the cerebral cortex with a sharp pair of scissors. Remove as much white matter as possible from the cortex by scraping with a blunt scalpel, weigh, then finely chop the tissue.

Homogenize the cortex in the isolation medium by six to eight complete strokes of the coaxial homogenizer attached to an electric motor rotating at 1500 rev/min.

Fractionate the homogenate by differential then density gradient centrifugation using the scheme outlined in Figs. 14.5 and 14.6.

Precautions   Good separation can be obtained using this scheme but it is essential to take the following precautions to obtain the optimum resolution and purification.

1. *Source of brain*   Obtain the pig brains from freshly killed animals and transport them from the slaughterhouse to the laboratory in a polythene bag placed on crushed ice in a Dewar flask.
2. *Gradients*   Precool all solutions to 0°C and carefully layer the solutions on top of each other. Leave the gradients for 2–4 h at 0–5°C before use.
3. *Washing of pellet*   It is essential to wash the mitochondrial pellet (P_2) at least three times before placing it on the Ficoll gradient.
4. *Suspension of pellet*   Suspend the pellet P_2 in 2 per cent w/v Ficoll by gently sucking the mixture up and down in a Pasteur pipette, then homogenize by one complete stroke in a homogenizer with a loose fitting pestle.
5. *Osmotic shock*   Rupture the synaptosomes by suspending them in glass distilled water (1 ml/g tissue) and forcing the suspension through a 13 gauge needle attached to a 25 ml syringe ten times. Transfer the suspension to a coaxial homogenizer and complete the suspension using five strokes by hand.

Experiment 14.8   The distribution of enzymes in the fractions obtained from pig brain

PRINCIPLE

The distribution of four enzymes among the subcellular fractions is investigated in terms of their total and relative specific activities.

Lactate dehydrogenase   This is a cytoplasmic enzyme marker, so little or no activity would be expected in the washed mitochondrial pellet. However, results show that LDH is present in bands B and D where it is occluded in synaptosomes and other membrane fragments.

Glutamate dehydrogenase   This enzyme is found in the mitochondrial matrix and would therefore be expected to occur in fraction F which is the mitochondrial pellet. However, some activity should also be detected in the synaptosomes (D) from the mitochondria present in these particles (Fig. 14.7). Osmotic lysis of these nerve-ending particles and subsequent fractionation can be used to isolate these synaptosomal mitochondria (K). Traces of GDH activity in other fractions are probably due to fragmented mitochondria.

Acetylcholinesterase   This enzyme is present on the post-synaptic membrane of cholinergic neurones (Fig. 14.7) and a high activity would be expected in the synaptosomes. However, separation and fragmentation of the post-synaptic membrane means that this activity is also found in other fractions.

$Na^+ - K^+$-*ATPase*   This enzyme is a marker for the plasma membrane and a high activity would be expected in the fraction containing the purified synaptosomal membranes (I).

The enzyme is assayed by linked enzyme system with the following sequence of reactions:

$$ATP \xrightarrow{\text{ATPase}} ADP + Pi$$

$$ADP + \text{phosphoenol pyruvate} \xrightarrow{\text{PK}} \text{pyruvate} + ATP$$

$$\text{pyruvate} + NADH + H^+ \xrightarrow{\text{LDH}} \text{lactate} + NAD^+$$

PK = pyruvate kinase

LDH = lactate dehydrogenase

In brain the optimal activity is found in the presence of 100 mmol/litre Na^+ and 30 mmol/litre K^+ and is inhibited by ouabain.

MATERIALS   10

1. Materials for the fractionation of pig brain (Exp. 14.7) — —
2. Reagents for the assay of glutamate dehydrogenase (Exp. 14.3b) — —
3. Reagents for the assay of lactate dehydrogenase (Exp. 14.3d) — —
4. Reagents for the assay of protein (Exp. 14.2) — —
5. Reagents for the assay of ATPase: — —

Tris–HCl buffer (100 mmol/litre, pH 7.4)	250 ml
Ionic solution (30 mmol/litre, $MgSO_4$; 1 mol/litre, NaCl; 200 mmol/litre, KCl)	50 ml
NADH (10 mg/ml in tris buffer)	10 ml
Pyruvate kinase	5 ml
Lactate dehydrogenase	2 ml
Phosphoenolpyruvate (0.1 mol/litre in water)	10 ml
KCN (1 mol/litre)	5 ml
Ouabain (10 mmol/litre)	25 ml
Na–ATP (0.1 mol/litre, pH 7.4)	25 ml

METHOD

Assay of Na$^+$ − K$^+$-ATPase   Prepare several test tubes containing the following reaction mixture and incubate at room temperature for 10 min:

	ml
Tris–HCl buffer	2.30
Ionic solution	0.30
NADH	0.05
Pyruvate kinase	0.03
Phosphoenolpyruvate	0.05
KCN	0.03
Water	0.09

Place the mixture in a UV spectrophotometer and establish the baseline by measuring the extinction at 340 nm for 1–2 min. At the end of this time, add 50 µl of suitably diluted sample and initiate the reaction by adding 100 µl of ATP solution. Once a good linear rate is established, add 100 µl of ouabain and measure the rate after the addition of the inhibitor. The Na$^+$ − K$^+$-ATPase is taken to be the difference between these two rates.

Assay of protein, GDH and LDH   Assay the protein content and the activity of these enzymes in each fraction using the methods described earlier (Exps 14.2, 14.3b, 14.3c).

Expression of results   Calculate the total activity of the enzymes in each fraction and determine their recoveries, then use the protein measurements to calculate the relative specific activities in each of the fractions. Express the results as shown earlier (Fig. 14.3) in terms of their distribution among the subcellular fractions and comment on the results.

Further reading

Dean, R. T., *Lysosomes.* Arnold, London, 1977.

Dingle, J. T., *Lysosomes a Laboratory Handbook*, 2nd edn. North Holland, Amsterdam, 1977.

Rickwood, D., *Centrifugation: A Practical Approach*, 2nd edn. IRL Press, Oxford, 1984.

Scheeler, P., *Centrifugation in Biology and Medical Science.* Wiley, New York, 1981.

Wrigglesworth, J. M., *Biochemical Research Techniques: A Practical Introduction.* Ellis Horwood, Chichester, 1983.

15. Photosynthesis and respiration

The oxygen electrode

Theory

The oxygen electrode is widely used in biochemical laboratories to monitor processes or reactions involving oxygen exchange. The evolution of oxygen by illuminated chloroplasts or the utilization of oxygen by respiring organisms and tissues can be readily followed with this equipment, although probably the commonest application of the technique is the recording of mitochondrial respiration.

The oxygen electrode consists of a platinum cathode and silver anode in saturated potassium chloride solution and, when a potential is applied across the 'cell' formed by these electrodes dipping in the test solution, oxygen is electrolytically reduced (Fig. 15.1). Four electrons are generated at the anode which are then used to reduce a molecule of oxygen at the cathode:

Anode $\quad 4Ag + 4Cl^- \longrightarrow 4AgCl + 4e$

Cathode $\quad 4H^+ + 4e + O_2 \longrightarrow 2H_2O$

$$4Ag + 4Cl^- \, 4H^+ + O_2 \longrightarrow 4AgCl + 2H_2O$$

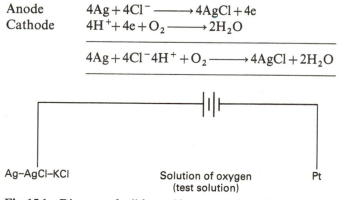

Ag–AgCl–KCl $\qquad\qquad$ Solution of oxygen $\qquad\qquad$ Pt
(test solution)

Fig. 15.1 Diagram of cell formed by oxygen electrode and test solution

If the polarizing voltage is in the range 0.5–0.8 V, then the current generated is proportional to the oxygen concentration in the medium. The amplified current is fed to a chart recorder which gives a trace of the change in oxygen concentration with time. Zero oxygen concentration is obtained by adding a crystal of sodium dithionite to the test solution and adjusting the pen to the baseline. The air-saturated buffer is taken to be 100 per cent oxygen and the pen adjusted accordingly. In practice, 100 per

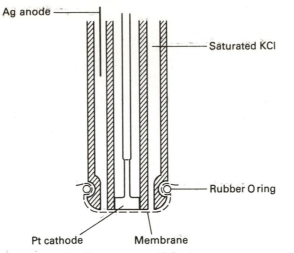

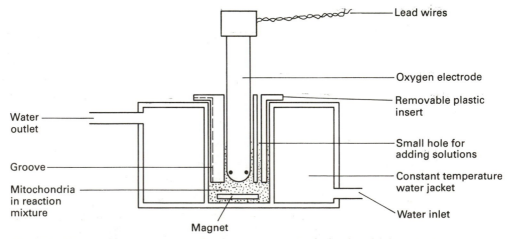

Fig. 15.2   The Clark oxygen electrode

cent oxygen is assumed to be 240 µmol of dissolved oxygen per litre, which is the solubility of oxygen in an aqueous solution at 26°C. If 4 ml is present in the reaction vessel then the total oxygen content is $4 \times 240 = 960$ nmol when saturated. For this reason, it is best to adjust the pen on the recorder to 96 rather than 100. In these circumstances, one division on the chart recorder is then equivalent to 10 nmol of O_2.

Practice

There are a number of electrode designs and the one illustrated in Fig. 15.2 is based on that of Clark. The electrode is separated from the bulk of the solution by a membrane to prevent deposition of materials on the electrode surfaces which would otherwise interfere with the oxygen determination (Fig. 15.2). The presence of a membrane does,

Fig. 15.3   The working arrangement for the measurement of mitochondrial respiration

however, reduce the time response of the electrode to changes in oxygen concentration and this can be a nuisance since meaningful values of oxygen consumption can only be obtained when the response time of the electrode is greater than the rate at which oxygen can be consumed or evolved. This problem can be overcome to some extent by keeping the solution well stirred. The electrode assembly is enclosed in a plastic vessel of about 4 ml capacity surrounded by a water jacket maintained at a constant temperature and reagents are added through a small hole in the top. There is therefore negligible diffusion of oxygen into the test solution from the surrounding air (Fig. 15.3).

One disadvantage of the Clark modification is that some compounds such as oligomycin become bound to the membrane and are very difficult to remove by washing. If this becomes a problem then the membrane has to be changed.

Experiments

Photosynthesis

Photosynthesis is the process by which light energy is used to synthesize carbohydrate in plants and algae from CO_2 and H_2O:

$$CO_2 + 2H_2O \longrightarrow (CH_2O) + O_2 + H_2O$$

The synthesis of carbohydrate takes place in two stages known as the light and dark reactions.

Light reaction   The light reaction only takes place in the presence of visible radiant energy which is absorbed by the green pigment chlorophyll present in chloroplasts. The light reaction consists essentially of the removal of electrons from water (photolysis) and these are then used to reduce $NADP^+$ and generate ATP. There are two light-driven reactions which take place at the reaction centres of photosystem I (PS I) and photosystem II (PS II) and operate in series. They consist of electron transport chains connected in the classical zigzag or Z scheme (Fig. 15.4).

Dark reaction   The second stage of photosynthesis does not require the presence of light and is therefore referred to as the dark reaction. This stage involves the utilization of NADPH and ATP generated by the light reaction to fix carbon dioxide. The complex metabolic sequence whereby CO_2 is effectively converted to sugars was elucidated by Calvin and his associates and is sometimes known as the Calvin cycle.

$$6CO_2 + 12NADPH + 18ATP^{4-} + 12H_2O$$
$$\downarrow$$
$$Glucose + 12NADP^+ + 18ADP^{3-} + 18P_i^{2-} + 6H^+$$

Experiment 15.1   The isolation of chloroplasts from spinach leaves

PRINCIPLE
The leaves are placed under strong illumination before homogenization as this reduces the starch grain content of the leaves and hence damage to the chloroplasts during isolation.

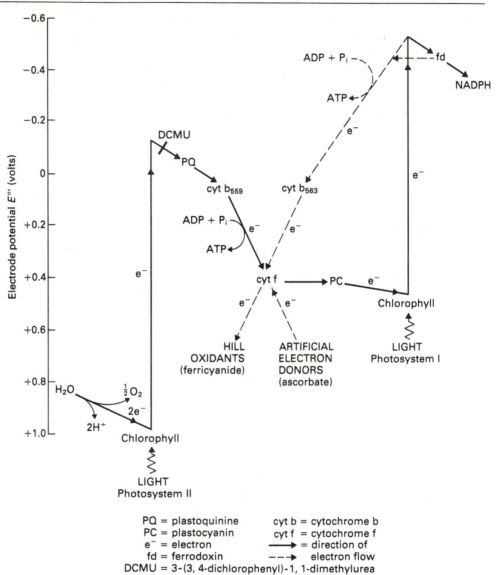

Fig. 15.4 Electron transport and the generation of NADPH and ATP in chloroplasts

The leaves are then homogenized in buffered hypertonic saline and the chloroplasts isolated by centrifugation.

MATERIALS $\underline{10}$
1. Fresh spinach leaves   500 g
2. Isolation medium (0.3 mol/litre NaCl, 3 mmol/litre $MgCl_2$,   500 ml
    0.2 mol/litre tricine, pH 7.6)

3. Double strength assay medium (0.2 mol/litre sorbitol,   500 ml
   6 mmol/litre $MgCl_2$, 0.4 mol/litre tricine, pH 7.6)
4. Waring blender and muslin   5
5. Centrifuge in cold room   5

METHOD

Remove the midribs from 100 g of spinach leaves and cut the leaves into small sections (approx 5 cm^2). Place the leaf material in ice water under strong illumination for about 1 h then homogenize with 100 ml of isolation medium in the Waring blender. Filter the suspension through six layers of muslin and centrifuge the filtrate at maximum speed in a bench centrifuge in the cold room for 30 s. Discard the supernatant and wash the pellet carefully with *single strength* assay medium spinning at maximum speed for 1 min. Resuspend the pellet in 2–3 ml of assay medium and store on ice until required.

Experiment 15.2   The evolution of oxygen by isolated chloroplasts using Hill oxidants

PRINCIPLE

Electrons flowing from PS II towards PS I or from ferredoxin towards $NADP^+$, can be intercepted by artificial electron acceptors. This process is known as the Hill reaction and can be used to measure the photochemical activity of PS I and PS I acting in series or of PS II alone. The photochemical activity of PS I can be measured if the PS II activity is blocked by the powerful herbicide dichlorodimethylurea (DCMU) and the electron flow provided by an artificial electron donor.

(*i*) *Ferricyanide*

$$2H_2O + 4Fe(CN)_6^{3-} \longrightarrow 4H^+ + 4Fe(CN)_6^{4-} + O_2$$

(*ii*) *2,6-Dichlorophenolindophenol (DCPIP)*

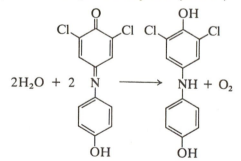

(*iii*) *Herbicide (DCMU)*

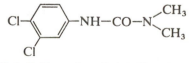

3-(3,4-dichlorophenyl)-1,1-dimethylurea

MATERIALS   10
1. Materials for the isolation of chloroplasts (Exp. 15.1)   —
2. Potassium ferricyanide (10 mmol/litre, prepared fresh)   50 ml
3. Dichlorophenolindophenol (1 mmol/litre, prepared fresh)   50 ml
4. Dichlorophenyldimethylurea (250 µmol/litre)   20 ml
5. Sodium dithionite (solid)   —
6. Oxygen electrodes   5

METHOD

Oxygen electrode assays   Set up and calibrate the oxygen electrode in the usual way and, after calibration, move the pen to the centre of the chart paper with the zero control as oxygen evolution is being measured.

   Start each experiment by adding 2 ml of double strength assay medium to the electrode and sufficient chloroplasts to give an apple-green suspension and a good rate of oxygen evolution in the experiment. This amount of chloroplasts should then be maintained throughout the investigation. Add sufficient distilled water and additions to give a final volume of 4 ml of single-strength assay medium.

   The assays should be set up in the dark by surrounding the electrode with aluminium foil; the rate of oxygen evolution is monitored in the light following the addition of cofactors, inhibitors, etc. The chloroplast suspension can be conveniently illuminated by a 100 watt bulb 'focused' roughly through a large round-bottomed flask filled with water.

Hill oxidants   Test the effectiveness of ferricyanide and DCPIP as Hill oxidants. Add increasing amounts of the oxidants recording both the rate and the total amount of O_2 evolved at each concentration. Plot the results as they are recorded.

   Finally, add sufficient DCPIP to ensure sustained O_2 evolution and test the effect of 50 µl of DCMU. What happens and why?

Experiment 15.3   Methyl viologen as a terminal electron acceptor

PRINCIPLE

Methyl viologen accepts electrons from PS I but the reduced compound is rapidly reoxidized:

$$H_2O + MV_{ox} \longrightarrow MV_{red} + \tfrac{1}{2}O_2$$
$$O_2 + MV_{red} \longrightarrow MV_{ox} + H_2O_2$$
$$\overline{\phantom{H_2O + MV_{ox} \longrightarrow MV_{red}}}$$
$$H_2O + O_2 \longrightarrow H_2O_2 + \tfrac{1}{2}O_2$$

Therefore 0.5 mol O_2 is taken up per pair of electrons transferred.

MATERIALS   10
1. Materials as for Exp. 15.2   —
2. Methyl viologen (10 mmol/litre, prepared fresh)   25 ml

3. Sodium azide (20 mmol/litre)   25 ml
4. Sodium ascorbate (50 mmol/litre, pH 7.6, prepared fresh)   25 ml

METHOD

Oxygen uptake with methyl viologen as electron acceptor   Measure the rates of oxygen uptake in the presence of methyl viologen with, and without, added azide (0.5 mmol/litre) to block catalase activity. Check the effects of varying the amount of methyl viologen and comment on the results.

Activity of PS I measured with methyl viologen   Measure the rate of oxygen evolution of the sample in the presence of DCPIP and let the reaction continue until the DCPIP is used up. Add DCMU to block PS II activity and carefully note what happens, then add sufficient sodium ascorbate to reduce the DCPIP so that the blue colour disappears. Measure the rate of oxygen uptake, add methyl viologen and compare the two rates of oxygen uptake. Repeat the experiment but this time omit the second DCPIP addition.

Comment on the results and discuss what is happening at each stage of the experiment.

Experiment 15.4   The spectrophotometric assay of the Hill reaction and the estimation of chlorophyll

PRINCIPLE

Hill reaction   DCPIP loses its blue colour on reduction and this can be followed in a spectrophotometer at 600 nm.

Assay of chlorophyll   The chlorophyll is extracted from the chloroplasts by shaking with acetone and the absorption of the solution is plotted in the visible region of the spectrum.

MATERIALS   10
1. Materials as in Exp. 15.2   —
2. Acetone (80 per cent v/v)   500 ml
3. Spectrophotometers   5

METHOD

Assay of Hill reaction   Instead of the oxygen electrode, prepare the sample in a spectrophotometer cuvette and measure the rate of DCPIP reduction by exposing the sample to successive 15 s or 30 s periods of illumination. Adjust the chlorophyll and DCPIP concentrations to yield a rate of reduction that can be conveniently measured.

Taking into account possible variations in the concentrations of the various components, how does the rate measured compare with the rate of oxygen evolution observed with DCPIP in the oxygen electrode?

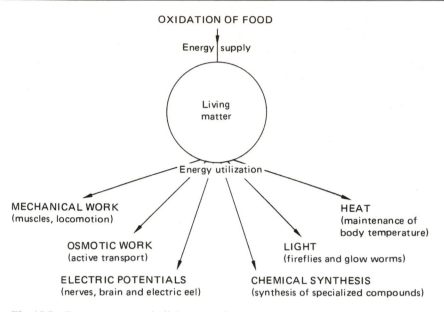

Fig. 15.5   Energy turnover in living organisms

Assay of chlorophyll content   Add a suitable volume of chloroplast suspension (0.2 ml) to 10 ml of the solvent, shake thoroughly and filter through Whatman No. 1 filter paper into a 25 ml volumetric flask. Rinse out the test tube with a further 5 ml of the aqueous acetone and use this to wash the filter paper. Repeat the washing and make up to 25 ml with the solvent.

Determine the extinction of the green solution against a solvent blank.

$$\text{Chlorophyll (mg/ml)} = \text{Extinction at 625 nm} \times 5.8$$

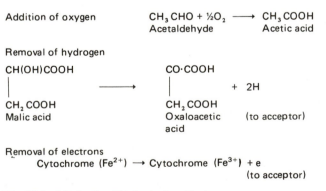

Fig. 15.6   Methods of biological oxidation

Oxidation and reduction in living cells

Living organisms require a continuous supply of energy to maintain the many and varied functions characteristic of living matter (Fig. 15.5). In most cases, this energy is obtained by the oxidation of metabolites from the digestion of food.

There are three ways that such oxidations can take place, although they all essentially involve the loss of electrons from the compound being oxidized (Fig. 15.6).

The chemical energy in food which is released during oxidation comes ultimately from the light energy of the sun during photosynthesis. The overall reaction in this process is one of reduction when electrons are added to the compound being formed.

Oxidation and reduction are always linked together rather like the formation of an acid and its conjugate base, you cannot have one without the other. It is therefore more accurate to talk about a *redox* reaction.

$$AH \rightleftharpoons A^- + H^+$$

Acid   Conjugate Proton
base

$$B \rightleftharpoons B^+ + e$$

Reductant Oxidant Electron

Electron transport chain

Nearly all biological oxidations take place by the removal of hydrogen from the substrate. Hydrogen atoms thus removed are then passed on to an acceptor, usually one of the pyridine nucleotides NAD^+ or $NADP^+$. In the aerobic cell, the reduction of NAD is followed by a series of electron transfers with the eventual formation of water and the generation of three molecules of ATP. This sequence of redox reactions is known as the *electron transport chain* or the *respiratory chain* (Fig 15.7) and like most other oxidative processes takes place in mitochondria.

Experiment 15.5   The respiration of mitochondria and oxidative phosphorylation

PRINCIPLE

When mitochondria are carefully isolated and suspended in an isotonic medium in the presence of substrate, only a slow rate of respiration is observed. On adding ADP, the respiration rate increases until all the ADP is phosphorylated, when the rate of respiration returns to the original slow rate. The quantity of ADP that is added is known and the amount of oxygen consumed is measured (X) so it is possible to arrive at a P/O ratio for the particular substrate used.

A typical trace of oxygen consumption with time is shown in Fig. 15.8 where A is the rate of respiration in the presence of substrate and B the rate of phosphorylating respiration. The ratio of these rates is known as the *respiratory control ratio* and is a measure of the degree of coupling of respiration and phosphorylation. A low ratio indicates loose coupling, but freshly prepared mitochondria should show a high respiratory control of 4 or more.

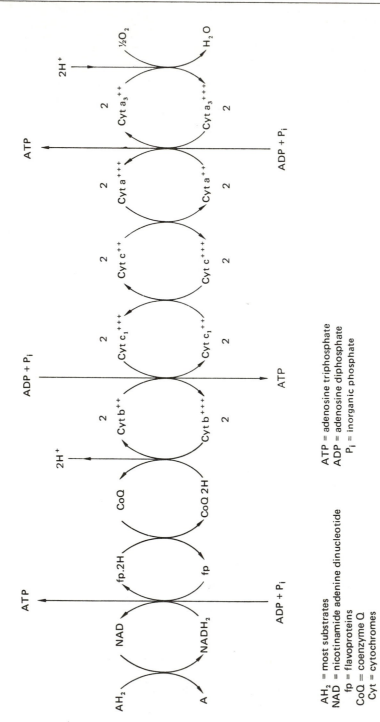

AH$_2$ = most substrates
NAD = nicotinamide adenine dinucleotide
fp = flavoproteins
CoQ = coenzyme Q
Cyt = cytochromes

ATP = adenosine triphosphate
ADP = adenosine diphosphate
P$_i$ = inorganic phosphate

Fig. 15.7   The electron transport chain

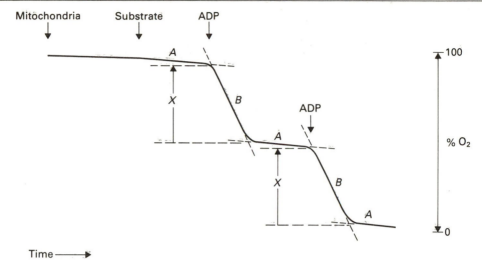

Fig. 15.8   A typical curve of oxygen uptake by mitochondria during substrate-induced and phosphorylating respiration

Calculations

$$\text{Respiratory control ratio} = \text{Rate } B/\text{Rate } A$$

$$\text{P/O ratio} = \mu\text{mol ADP added}/\mu\text{atoms O}_2 \text{ utilized } (X)$$

The phosphorylation sites in the electron transport chain are shown in Figs. 15.7 and 15.9, so that substrates using $NADH_2$ as coenzyme should give a P/O ratio of 3 while succinate, which bypasses the first phosphorylation site, should give a ratio of 2. The mixture of ascorbate and tetramethylphenylenediamine (TMPD) uses cytochrome c as electron acceptor and thereby misses two of the phosphorylation sites so that the P/O ratio for this substrate is 1.

Even in freshly prepared mitochondria, the P/O ratios actually found are always less than the whole number values given above due to partial uncoupling and the action of membrane ATPases.

MATERIALS   <u>10</u>
1. Materials for the isolation of rat liver mitochondria (Exp. 14.4)   —
2. Isolation medium (0.3 mol/litre sucrose; 2.5 mmol/litre tris–HCl, pH   2 litres
   7.4; 0.5 mmol/litre EDTA)
3. Incubation medium: 500 ml
     Sucrose (150 mmol/litre)
     Potassium chloride (20 mmol/litre)
     Magnesium chloride (20 mmol/litre)
     Potassium phosphate (1 mmol/litre, pH 7.4)

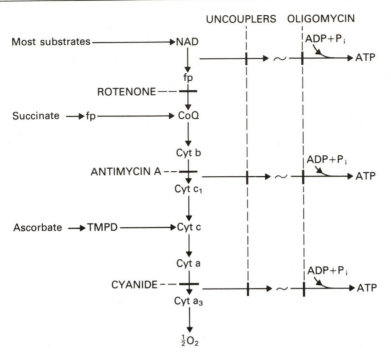

~ Proton gradient
Inhibitors shown in capitals (e.g., CYANIDE)

Fig. 15.9   The site of action of inhibitors and uncouplers of the respiratory chain

4. Sodium malate (200 mmol/litre, pH 7.4)   25 ml
5. Sodium succinate (200 mmol/litre, pH 7.4)   25 ml
6. Ascorbate–TMPD (200 mmol/litre ascorbate, pH 7.4 containing   25 ml
   5 mmol/litre tetramethylparaphenylenediamine prepared fresh and
   stored in the dark)
7. ADP (20 mmol/litre, pH 7.4, prepared fresh)   10 ml

METHOD

As with previous experiments using the oxygen electrode, make any additions to the
mitochondrial suspension in small volumes only (20–100 μl) so that the total volume
in the electrode compartment remains close to 4 ml.

Prepare a fresh suspension of rat liver mitochondria as described in Exp. 14.4 and
store on ice until required. Set up and calibrate the oxygen electrode as described
earlier and add a small volume of the mitochondrial suspension (50–100 μl) followed
by succinate and ADP in that order. Calculate the P/O ratio and the respiration

control index; repeat the experiment using malate and ascorbate/TMPD as the substrate.

Experiment 15.6   The effect of inhibitors on the respiratory chain

PRINCIPLE

Many compounds are known to inhibit electron flow in the respiratory chain and the sites at which they act can be identified by examining the effect of these inhibitors on respiration induced by the substrates considered in the previous experiment.

The three inhibitors examined are rotenone, antimycin A and cyanide. Rotenone blocks electron flow between the nicotinamide nucleotides and flavoproteins; antimycin A acts at the cytochrome b/cytochrome c step; and cyanide inhibits cytochrome oxidase at the end of the respiratory chain (Fig. 15.9).

All three inhibitors are highly toxic and must be treated with extreme care.

MATERIALS	10
1. Materials as in Exp. 15.5	—
2. Rotenone (1 mmol/litre in 95 per cent v/v ethanol)	10 ml
3. Antimycin A (0.1 mg/ml in 95 per cent v/v ethanol)	10 ml
4. Potassium cyanide (10 mmol/litre)	10 ml

METHOD

Repeat the previous experiment, then examine the effect on the oxygen uptake of adding a small volume (10–50 μl) of one of the inhibitor solutions. Take each of the three substrates in turn and record the changes in mitochondrial respiration induced by each of these inhibitors. What conclusions can you draw about the site of action of these inhibitors?

Experiment 15.7   Compounds that affect the high energy state of mitochondria

PRINCIPLE

The rate of respiration in undamaged mitochondria is well below the maximum possible but in the presence of uncouplers, this restriction is removed and electron flow is stimulated.

The electron transport chain sets up a proton gradient across the mitochondrial inner membrane and it is the discharge of this gradient via a reversed ATPase that generates the formation of ATP. 2,4-DNP discharges the proton gradient across the membrane so that the two processes become *uncoupled*. Respiration continues but ATP is no longer produced. The practical effect of an *uncoupling agent* is to cause an increase in the rate of respiration and loss of respiratory control.

Oligomycin does not discharge the proton gradient but inhibits the mitochondrial ATPase so that ATP cannot be formed. Oligomycin therefore inhibits the electron flow indirectly by preventing the discharge of the proton gradient by the ATPase. As expected, this inhibition can be released by 2,4-DNP.

MATERIALS   <u>10</u>
1. Materials as for Exp. 15.5   —
2. 2,4-Dinitrophenol (5 mmol/litre in ethanol)   10 ml
3. Sodium salicylate (0.2 mol/litre)   10 ml
4. Oligomycin (0.5 mg/ml)   10 ml

METHOD

Measure the respiration of freshly prepared mitochondria from rat liver, as described in Exp. 15.5, then examine the effect on the respiration of adding 20–50 µl of each of the above solutions to the reaction mixture. What effect do these compounds have on substrate-induced and phosphorylating respiration? How does this alter the respiratory control?

Experiment 15.8   Calcium transport in mitochondria

PRINCIPLE

Mitochondria will accumulate Ca^{2+} if they are provided with an energy source. This energy-linked uptake of Ca^{2+} is supported by the membrane potential produced either by respiration or by the hydrolysis of ATP.

Ca^{2+} uptake can be measured in various ways using either direct methods, such as the accompanying increase in oxygen consumption or release of protons, or indirect methods, such as ^{45}Ca, a Ca^{2+} specific electrode, or a Ca^{2+} specific indicator (murexide). In this experiment, the Ca^{2+} transport system in mitochondria is followed by the increase in oxygen consumption.

Carbonylcyanide-4-trifluoromethoxyphenylhydrazone (FCCP) is an uncoupling agent while ruthenium red and lanthanum chloride are inhibitors of Ca^{2+} transport. The effect of the other inhibitors on mitochondria has been described already (Exp. 15.6).

MATERIALS   <u>10</u>
1. Materials for the respiration of rat liver mitochondria (Exp. 15.5)   —
2. Isolation medium (0.3 mol/litre sucrose; 2.5 mmol/litre tris–HCl   2 litres
   buffer, pH 7.4)
3. Reaction medium (120 mmol/litre KCl; 5 mmol/litre tris–HCl   500 ml
   buffer, pH 7.4)
4. Substrates (prepared fresh)   25 ml
      Potassium succinate (0.5 mol/litre, pH 7.4)
      Potassium pyruvate–malate (4:2) (0.5 mol/litre, pH 7.4)
      Potassium ascorbate–TMPD (0.2 mol/litre–10 mmol/litre,
      pH 7.4)
5. Ca^{2+} transport reagents   25 ml
      Calcium chloride (40 mmol/litre)
      Lanthanum chloride (1 mmol/litre)
      Ruthenium red (1 mmol/litre)

6. Inhibitors and uncouplers   25 ml
    Rotenone (1 mg/ml in 95 per cent v/v ethanol)
    Antimycin A (1 mg/ml in 95 per cent v/v ethanol)
    Oligomycin (1 mg/ml in 95 per cent v/v ethanol)
    Carbonyl cyanide p-trifluoromethoxyphenylenehydrazone
    (FCCP, 1 mmol/litre in 95 per cent v/v ethanol)
    Potassium cyanide (10 mmol/litre)

METHOD

Prepare mitochondria from rat liver as previously described using the 0.3 mol/litre buffered sucrose and check that they are coupled. Determine the mitochondrial protein content using the deoxycholate–biuret method (Exp. 14.2), then carry out the following experiments with the appropriate controls.

1. Add mitochondria to the reaction medium in the electrode cuvette to give a final concentration of about 2.5 mg of protein per ml and a volume of 3.9 ml. Add succinate as substrate to a final concentration of 5 mmol/litre. When the respiration is steady, add 50 nmol Ca^{2+} per mg of protein. Make further additions of Ca^{2+} until no further response is obtained.
2. Repeat (1) in the presence of a permeant anion (2 mmol/litre phosphate, acetate or oxalate). Are the results different? If so why?
3. Repeat (1) using different respiratory substrates and calculate (i) the rates of Ca^{2+}-induced respiration and (ii) the Ca^{2+}/O ratio, assuming all the Ca^{2+} added was taken up.
4. Investigate the effects of the inhibitors and uncouplers on the Ca^{2+} uptake.
5. Investigate the effects of the Ca^{2+}-transport inhibitors ruthenium red and lanthanum. What concentrations are needed to cause 50 per cent inhibition?

Further reading

Hall, D. O. and Rao, K. K., *Photosynthesis*, 3rd edn. Arnold, London, 1981.
Halliwell, B., *Chloroplast Metabolism*, Clarendon Press, Oxford, 1984.
Nicholls, D. G., *Bioenergetics—An Introduction to the Chemiosmotic Theory*. Academic Press, London, 1982.
Tzagaloff, A., *Mitochondria*, Plenum Press. New York, 1982.

16. Molecular biology

Definition

Molecular biology is a fashionable term and occurs frequently in the scientific literature. However, the exact meaning is not always clear and the expression molecular biology has been used in a variety of ways.

Molecular biology in the broadest sense can be used to describe the study of biology at the molecular level. This definition is too general and in any case there is already an adequate definition of this area of knowledge, namely biochemistry.

The expression molecular biology was first used by workers in the field of protein structure and by the 1950s had come to mean the study of macromolecules and their properties. The discovery of the structure of DNA in 1953 had a major impact on biology and since then the term molecular biology has been used to describe the study of gene structure, gene replication and gene expression. Molecular biology therefore involves the detailed study of the events summarized below.

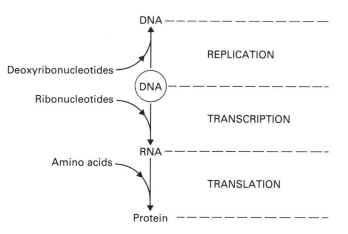

Relationship to biochemistry

Molecular biology can be considered as a discipline at the interface of genetics and biochemistry so that:

$$\text{Genetics} + \text{Biochemistry} = \text{Molecular biology}$$

Another view is that molecular biology is a way of defining an area of biochemistry, rather like enzymology, immunochemistry or intermediary metabolism. In any case, molecular biology and biochemistry are inseparable and a good knowledge of biochemistry is required to understand the subject.

Experiments

Molecular biology is a rapidly expanding area of biochemistry and the following experiments are only a small selection from this field of study. Many experiments are unsuitable for class work as they require a good grasp of technical and manipulative skills which can only really be learned by working on a project in this area. The following experiments have been selected for class work but they need to be carried out carefully and skilfully if good results are to be obtained.

Other molecular biology experiments can be found in the section on nucleic acids (Chapter 11).

Experiment 16.1   Induction of β-galactosidase in strains of *Escherichia coli* (i^+ and i^-)

PRINCIPLE

Enzyme induction   Bacteria are very adaptable organisms and an alteration in the nutritional environment of some cells leads to a change in the enzyme production, enabling the cells to grow under the new conditions. For example, some *Escherichia*

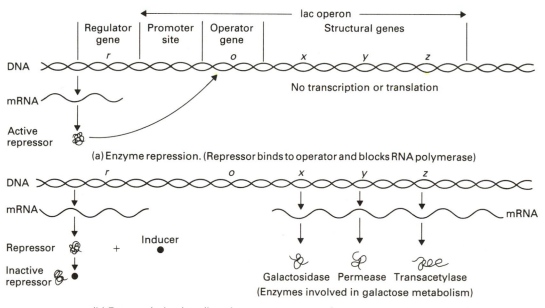

Fig. 16.1   The regulation of the lac operon in *E. coli*

coli react to form enzymes associated with the metabolism of β-galactosides when such compounds are introduced into the growth medium. Enzymes that are rapidly synthesized in increased amounts in response to the substrate in the medium are known as *inducible enzymes*. This is in contrast to the *constitutive enzymes* which are found in constant amounts irrespective of the metabolic state of the cell.

lac Operon　In prokaryotic cells such as *E. coli* the cellular DNA is contained in a single circular chromosome which contains genes for the production of mRNA (structural genes) and genes associated with transcriptional control (control genes), so that the structural genes are only transcribed when the control genes allow transcription to occur. The mechanism for this, known as the *operon model*, was proposed by Jacob and Monod and is illustrated in Fig. 16.1.

The lac operon is the gene cluster responsible for the regulation of the three enzymes needed for the metabolism of galactose. It consists of three structural genes and a control region that contains a *promoter* and *operator* site. RNA polymerase binds to the promoter site and this is enhanced by a cyclic AMP receptor protein (CAP) which also binds in this region. Immediately adjacent to this area is the operator which is the binding site for a *repressor* coded by the regulatory gene. In the absence of an inducer, the repressor binds to the operator site and blocks RNA polymerase so that the structural genes cannot be expressed (Fig. 16.1(a)). However, when the inducer is present, it binds to the repressor protein converting it to an inactive form that cannot bind to the operator site, so that the lac operon can be transcribed (Fig. 16.1(b)).

Bacterial strains　In this experiment, two strains of *E. coli* are investigated. In the *wild type* ($i^+ x^+ y^+ z^+$), all the proteins are inducible and are only synthesized when an inducer is present. In contrast, the *constitutive mutant* ($i^- x^+ y^+ z^+$) contains high levels of the three enzymes in the absence of an inducer.

Enzyme assay　The substrate used to measure the β-galactosidase activity is *o*-nitrophenyl-β-D-galactoside which is hydrolysed by the enzyme to *o*-nitrophenol and galactose.

The activity is assayed colorimetrically by adding alkali which stops the reaction and develops the full yellow colour of *o*-nitrophenol.

Laboratory safety　Use only cotton wool plugged pipettes for pipetting *E. coli* and put all glassware that contained bacteria into buckets containing lysol or a similar antiseptic.

MATERIALS　　　　　　　　　　　　　　　　　　　　　　　　　　10
  1. *Escherichia coli* (wild type)　　　　　　　　　　　　　　　　—
  2. *Escherichia coli* (constitutive mutant). Strain ML308 NCIB 9553,　—
     ACTC 15224

3. Bacterial growth medium. (Buffered salts medium containing a   10 litre
   carbon source (10 g/litre sodium succinate), amino acids for
   protein synthesis (1 g/litre vitamin-free casamino acids), and
   thiamine (10 mg/litre). The inducer or inhibitor is added to the
   medium as appropriate
4. Cetyltrimethylammonium bromide (0.8 mg/ml)   100 ml
5. *o*-Nitrophenyl-*β*-D-galactoside (1 mg/ml in buffer prepared just   200 ml
   before use)
6. Na_2CO_3 (1 mol/litre)   200 ml
7. Standard *o*-nitrophenol (100 µmol/litre in 0.1 mol/litre Na_2CO_3) 100 ml
8. Sterile pipettes plugged with cotton wool   —
9. Lysol or similar solution for sterilizing glassware after use   —
10. Colorimeter   5
11. Equipment for the large-scale growth of *E. coli*   —
12. Incubator or constant temperature room at 30°C   —
13. Lactose (200 mmol/litre)   100 ml
14. Chloramphenicol (1 mg/ml)   100 ml

METHOD

Growth of bacteria   Grow the two strains of bacteria overnight in the growth
medium, but in the absence of any inducer. Suspend the organisms in 5 ml final
volume of growth medium at a concentration of 5 mg/ml and set up five test tubes
containing 4 ml of growth medium for each strain of *E. coli*. Add 0.5 ml of the bacterial
suspension, warm to 30°C and at zero time add 0.5 ml of lactose (200 mmol/litre) as
inducer.

Stop one tube of each strain after incubation for 0, 30, 60, 90 and 120 min by adding
0.5 ml of chloramphenicol. Include a control by incubating each strain for 120 min in
the absence of inducer. Centrifuge each tube, wash the pellet once with growth
medium to remove inducer, and resuspend in 5 ml of growth medium and relate the
number of bacteria to the turbidity. Determine the cell growth by measuring the
turbidity at 600 nm and assay the *β*-galactosidase activity in each sample.

What is the time course of induction of *β*-galactosidase? From your results deduce a
suitable incubation time for subsequent experiments.

Assay of β-galactosidase   This enzyme is assayed in disrupted cells because transport
across the cell membrane may be rate-limiting in the intact cells. The cell membrane is
disrupted without loss of activity by adding the detergent cetyltrimethylammonium
bromide at a final concentration of 0.1 mg/ml.

Shake 1 ml of the washed bacterial suspension in a centrifuge tube with 0.5 ml of
cetyltrimethylammonium bromide (0.8 mg/ml). Add 1.5 ml of freshly prepared *o*-
nitrophenyl-*β*-D-galactoside and incubate at 30°C for 5 min or longer if necessary.
Include a control with 1 ml of growth medium but no organisms. Stop the reaction by
adding 1 ml of Na_2CO_3 (1 mol/litre) and centrifuge to remove the bacteria. Decant the

supernatant and determine its *o*-nitrophenol concentration by comparing its absorbance at 420 nm with a standard curve of *o*-nitrophenol (0–100 µmol/litre) prepared in 0.1 mol/litre Na_2CO_3.

Express the activity as nanomoles of *o*-nitrophenyl-β-D-galactoside hydrolysed per minute per milligram cell wt.

Experiment 16.2   The effect of different inducers on the induction of β-galactosidase

PRINCIPLE

In some cases β-galactoside analogues may stimulate induction but they do not act as substrates for the induced enzyme. This non-metabolizable induction is called *gratuitous induction*. An example of a gratuitous inducer is β-thiogalactoside. Gratuitous inducers enable induction to be studied without interference from substrate or products but they may competitively inhibit β-galactosidase (Fig. 16.2).

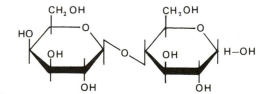

D-Galactose   Thiomethyl β-D-galactopyranoside

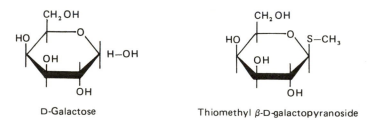

Lactose (β-D-galactosyl-1,4-D-glucose)

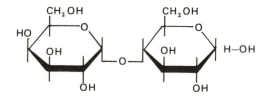

Melibiose (α-D-galactosyl-1,4-D-glucose)

Fig. 16.2   Inducers of β-galactosidase activity in *E. coli*

MATERIALS AND METHODS

Using the previous experiment as a model, investigate, with controls, the efficiency of the following compounds as inducers of β-galactosidase. Use each compound at several concentrations and try to estimate the concentration of inducer that gives half maximal stimulation.

1. Thiomethyl β-D-galactopyranoside (up to 10 mmol/litre).
2. Melibiose (0.5 mol/litre).
3. Lactose (0.5 mol/litre).
4. Galactose (0.5 mol/litre).

Stock solutions of the above compounds should be prepared fresh; 50–100 ml should be adequate for five pairs of investigators.

Experiment 16.3   The effect of protein synthesis inhibitors on the induction of β-galactosidase

PRINCIPLE

Chloramphenicol inhibits protein synthesis by binding to the 70 S ribosomes found in prokaryotic cells and mitochondria and blocking peptidyl transferase. However, this antibiotic does not affect protein synthesis by the 80 S ribosomes of eukaryotic cells.

Cycloheximide on the other hand prevents peptide bond formation by binding to the 80 S ribosomes of eukaryotes but is without effect on the 70 S ribosomes of prokaryotes.

MATERIALS AND METHOD

Investigate the effects of chloramphenicol and cycloheximide on the induction of the enzyme by lactose using concentrations of inhibitor up to 1 mg/ml.

Experiment 16.4   Turnover of β-galactosidase

METHOD

Induce the *E. coli* for 60 min with lactose as in previous experiments; then add chloramphenicol and follow the rate of decay of β-galactosidase activity over the next 2 h. Include a control with chloramphenicol added before the lactose.

What is the half-life of the enzyme inside the cell?

Experiment 16.5   Reticulocytes and their use in the study of protein synthesis

PRINCIPLE

Reticulocytes are immature red blood cells which contain the normal protein synthesizing machinery typical of eukaryotic cells. They are produced in bone marrow and appear in the blood in large numbers in response to phenylhydrazine-induced anaemia. The intact cells will readily incorporate amino acids from a suitable medium and, if the amino acids are tagged with a radioactive atom, protein synthesis can be monitored. As with most eukaryotic cells, the mRNA molecules found in reticulocytes are relatively stable compared to their bacterial counterparts. Thus this cell, whose protein synthesizing machinery is mainly geared to the synthesis of large amounts of globin, yields a lysate which contains globin mRNA (species for α and β chains are present) in addition to ribosomes, tRNA and other factors. Also present, but in small amounts, are mRNA species for proteins other than globin. Again protein synthesis may be studied by adding radioactive amino acids to the lysate. The translation process can therefore be studied *in vitro* using the natural endogenous messenger rather than employing exogenous templates.

The reticulocyte lysate system is readily available and has also been widely used for assessing the template activity of exogenous mRNA. However, in this case the very presence of globin mRNA creates a problem, as a high background incorporation of amino acids into globin occurs. Therefore, a measure of overall protein synthesis in terms of incorporation of labelled amino acids into total protein cannot be used as an activity index for added mRNA although this problem can be overcome if product analysis is performed by antiserum precipitation or SDS electrophoresis. An alternative approach is to try and eliminate the endogenous mRNA using ribonuclease, without damaging other types of RNA. Exogenous mRNA is then added as the sole template for translation.

During maturation *in vivo* reticulocytes lose their ability to initiate globin chains and eventually degrade all mRNA and ribosomes. Mature animal erythrocytes are therefore incapable of protein synthesis, although this ability is retained by avian erythrocytes.

MATERIALS

	10
1. Rabbits	5
2. Phenylhydrazine hydrochloride (2.5 per cent w/v adjusted to pH 7 with dilute NaOH)	50 ml
3. Wash medium (0.14 mol/litre NaCl, 0.05 mol/litre KCl, 5 mmol/litre Mg acetate)	250 ml
4. Sterile glassware and centrifuge tubes	—
5. Syringes and needles for blood collections	10

METHOD

Reticulocytosis is induced in rabbits by subcutaneous injections of a 2.5 per cent w/v solution of phenylhydrazine hydrochloride solution using the following protocol: day 1, 1.0 ml; day 2, 1.0 ml; day 3, 1.0 ml; day 4, 1.0 ml; day 5, 0.5 ml.

Collect 5–10 ml of blood from an ear on day 9 when a reticulocyte content of more than 80 per cent should be obtained. Centrifuge the blood on a bench centrifuge for 7 min at 4°C. Remove the supernatant with a Pasteur pipette (autoclaved), taking care to remove any buffy coat which appears at the interface between the packed cells and the plasma. Resuspend the cells in the wash medium with a glass rod and centrifuge again. Repeat the washing procedure twice more, then give a final spin for 15 min.

Discard the supernatant and store the cells on ice until required. If a lysate is required then retain 0.5 ml of the cells and lyse the remainder as detailed later.

Experiment 16.6   Protein synthesis in intact rabbit reticulocytes

PRINCIPLE

The value and use of reticulocytes is discussed in Exp. 16.5.

MATERIALS				10
1. Rabbit reticulocytes (Exp. 16.5)				—
2. Ferrous ammonium sulphate (0.105 per cent w/v, prepare fresh)				50 ml
3. Glass fibre discs (Whatman SF/A 2.1 cm)				—
4. Scintillation vials				—
5. Trichloracetic acid (5 per cent w/v)				250 ml
6. Trichloracetic acid (10 per cent w/v)				250 ml
7. Water baths at 80°C and 37°C (shaking)				5
8. Ethanol:acetone (1:1)				200 ml
9. Ovens at 50°C				5
10. Scintillation fluid				—
11. Liquid scintillation counter				1
12. Saline solution (0.13 mol/litre NaCl, 0.005 mol/litre KCl, sterile)				100 ml

13. Amino acid mixture. (The mixture is prepared by weighing out the
following mg of each amino acid and dissolving in 100 ml of the
saline solution.)

Alanine	23.0	Glycine	50.0	Methionine	6.0
Arginine HCl	16.0	Histidine	63.0	Phenylalanine	33.0
Aspartic acid	48.0	Isoleucine	5.0	Proline	20.0
Cysteine HCl	6.0	Leucine	33.0	Serine	22.0
Glutamine	50.0	Lysine HCl	41.0	Threonine	25.0
Tryptophan	8.0	Tyrosine	19.0	Valine	46.0

14. Master mix   —

	ml
Amino acid mixture	2.70
$MgCl_2$ (0.25 mol/litre in 10 per cent w/v glucose)	0.15
Tris–HCl (0.15 mol/litre, pH 7.6)	1.45
Trisodium citrate (0.01 mol/litre)	1.10
Sodium bicarbonate (0.01 mol/litre prepared fresh)	1.60
Total volume	7.00

15. Radioactive master mix. (Dissolve leucine in the master mix to give a   —
final concentration of 1 mmol/litre and add 20 μl of ^{14}C leucine with
an activity of 50 μCi/ml)

METHOD

Protein synthesis   Mix 0.25 ml of washed reticulocytes with 0.75 ml of radioactive
master mix, add 3.75 μl/ml of freshly prepared ferrous ammonium sulphate solution

and incubate the mixture at 37°C in a shaking water bath. Remove duplicate 50 μl aliquots at 0, 10, 20, 30, 40 and 60 min (gently mix before sampling), then spot each sample on to a glass fibre disc which is then placed in a scintillation vial. Cover the discs with 4 ml of 5 per cent w/v TCA. When all the samples are collected, wash the discs as follows:

1. 5 ml 10 per cent w/v TCA for 10 min;
2. 5 ml 10 per cent w/v TCA at 80°C for 10 min;
3. 5 ml ethanol:acetone for 5 min.

Please note that the supernatants will be radioactive so remove with a Pasteur pipette into an appropriate container.

Dry in an oven at 50°C for 10–15 min, add scintillation fluid and count the radioactivity in a scintillation counter.

Calculations
1. Determine the specific activity of ^{14}C leucine in the incubation mixture.
2. Plot the counts per minute as a function of time and comment on the shape of the graph.
3. Determine the rate of globin synthesis (pmol min^{-1} ml^{-1} packed cells) assuming 17 leucine residues per globin chain.

Note: 1 μCi $= 2.22 \times 10^6$ disintegrations per minute (d.p.m.).

Experiment 16.7   Protein synthesis in reticulocyte lysates

PRINCIPLE
The use of reticulocyte lysates to study protein synthesis is discussed in Exp. 16.5.

MATERIALS	10
1. Rabbit reticulocytes (Exp. 16.5)	—
2. Materials for Exp. 16.6 (items 1–11)	—
3. Sterile distilled water	—
4. Sterile glassware and pipette tips	—
5. Deep freeze at -70°C	1
6. Cocktail (5 × final strength)	10 ml
Tris–HCl, 50 mmol/litre, pH 7.6	
KCl, 375 mmol/litre	
Mg acetate, 10 mmol/litre	
ATP, 5 mmol/litre	
GTP, 1 mmol/litre	
glucose, 15 mmol/litre	
Check the final pH and adjust with KOH if necessary	
7. Radioactive cocktail	—
675 μl cocktail × 5	
10 μl ^3H leucine (1 μCi per μl)	

        10 µl transfer RNA (7 mg in 2 ml)

        50 µl amino acid mix (The mixture contains
        a range of amino acids less leucine,
        available from Amersham International)

8.	Energy mix (creatine phosphate, 23 mg/ml + creatine phosphokinase, 10 mg/ml. Prepare fresh and keep on ice)	1 ml
9.	Bovine serum albumin (100 mg/ml)	1 ml
10.	Whatman No. 1 filter paper discs (2.1 cm diameter)	—
11.	Ethanol (95 per cent v/v)	100 ml
12.	Water baths at 30°C.	5

METHOD

Preparation of lysate   Prepare washed reticulocytes as described in Exp. 16.5. Measure the packed cell volume and lyse the cells by adding an equal volume of ice-cold sterile distilled water. Centrifuge at $5000\,g$ for 20 min at 4°C and remove the supernatant (lysate). Use the lysate immediately or freeze aliquots of the supernatant rapidly in sterile vials that have been pre-chilled to $-70°C$ for 1 h.

Protein synthesis   Thaw the lysate tube in the hand but do not allow to rise above 4°C, then store in an ice bath. Prepare the following incubation mixture in 15 ml glass centrifuge tubes which have been autoclaved

    40 µl radioactive cocktail
    15 µl energy mix
    10 µl BSA
    15 µl water or other addition

Add 70 µl of lysate to initiate the reaction and incubate at 30°C. Withdraw 10 µl aliquots for treatment and counting at 0, 2, 5, 10, 15 and 30 min in duplicate.

Treatment and counting of labelled aliquots   Mark in pencil a suitable number of filter paper discs, spot 10 µl aliquots on the discs and allow to dry for about 3 min. Place the discs in a beaker of cold 5 per cent w/v TCA and leave for 10 min. Repeat with hot (80°C) TCA then repeat with cold TCA for 10 min. Finally, wash with 95 per cent v/v ethanol to remove the TCA. This process must be carried out with care in view of the small volumes involved and so the discs should be held in tweezers until the aliquots are dry; do not place on the bench.

    Place the filter in insert tubes and dry in an oven at 50°C for 10–15 min, then count the radioactivity in a toluene-based scintillator.

CALCULATIONS
1. Plot graphs showing the incorporation of radioactivity into protein with time and account for the shape of the progress curves.
2. Determine the per cent c.p.m. incorporated into globin over the 30 min time period (i.e., per cent of c.p.m. added).

3. Determine the nanomoles of protein produced over the 30 min time period given that the molecular weight of globin is 64 000.
4. Is it possible to compare the efficiency of protein synthesis in the lysate with that of the intact cells?

Experiment 16.8   Compounds that affect the rate of protein synthesis in reticulocyte lysates

MATERIALS	10
1. Materials as for Exp. 16.7	—
2. Haemin solution (0.5 mmol/litre). Dissolve the haemin in 0.25 ml of sterile 1 mol/litre KOH, then add the following components in order, mixing thoroughly after each addition.	10 ml

  0.5 ml, 0.2 mol/litre, tris–HCl, pH 7.6
  8.9 ml, ethylene glycol
  0.19 ml, HCl, 1 mol/litre
  0.16 ml, water
  Store at −20°C in 100 µl aliquots until required

3. Chloramphenicol (5 mmol/litre)	10 ml
4. Cycloheximide (5 mmol/litre)	10 ml
5. Puromycin (5 mmol/litre)	10 ml

METHOD

Haemin   Using the reticulocyte lysate, devise an experiment to determine the effect of haemin (50 µmol/litre final concentration) on globin synthesis, bearing in mind that the effect of haemin may be time dependent.

Inhibitors of protein synthesis   Examine the effect of cycloheximide, chloramphenicol and puromycin at a final concentration of 1 mmol/litre on protein synthesis by the reticulocyte lysate.

Experiment 16.9   The isolation of bacterial DNA

PRINCIPLE

The bacteria are first digested with muramidase (lysozyme) in order to weaken the cell walls then ruptured with the detergent sodium dodecyl sulphate. The high salt concentration (0.15 mol/litre NaCl) in the extraction medium helps to prevent strand separation by providing a counter-ion shield round the double helix of DNA. EDTA is also present and this chelates metal ions needed for DNase and so inhibits the activity of this enzyme. The citrate present in the standard saline citrate (SSC) used later performs the same function.

Protein is denatured by the treatment with the chaotropic agent $NaClO_4$ and the extraction is carried out with the organic solvent mixture. Chloroform is the most important of the solvents and the small amount of isoamyl alcohol is present as an antifoaming agent.

After removal of the protein, DNA is purified by precipitation with ethanol then isopropanol.

The incubation with RNase should ensure that contamination of the DNA by RNA is kept to a minimum.

MATERIALS

		10
1.	*E. coli* or other organisms (3 g of packed cells per pair of students)	15 g
2.	Saline–EDTA (0.15 mol/litre NaCl, 0.1 mol/litre EDTA adjusted to pH 8.0)	250 ml
3.	Water baths at 37°C	5
4.	Muramidase (lysozyme)	100 mg
5.	Water baths at 60°C	5
6.	Sodium dodecyl sulphate (10 per cent w/v)	50 ml
7.	Sodium chlorate (6 mol/litre)	100 ml
8.	Chloroform:isoamyl alcohol (24:1)	500 ml
9.	Ethanol	250 ml
10.	Concentrated saline citrate (10 × SSC), (1.5 mol/litre NaCl, 0.15 mol/litre Na citrate, pH 7.0). This is diluted × 10 to give standard saline citrate (SSC) or × 100 to give 0.1 SSC as required	100 ml
11.	RNase (5 mg/ml, heated to 100°C for 2 min then cooled to inactivate any contaminating DNase	2 ml
12.	Sodium acetate (3 mol/litre)	25 ml
13.	Isopropanol	100 ml
14.	Reagents for the assay of DNA (Exp. 11.8)	—
15.	Reagents for the assay of RNA (Exp. 11.9)	—
16.	UV spectrophotometers	5
17.	NaOH (6 mol/litre)	50 ml
18.	Reagents for Exp. 11.7	—
19.	Reagents for Exp. 11.6	—

METHOD

Extraction   Mix 3 g of packed cells with 30 ml of the saline–EDTA solution in a 250 ml flask ensuring that the cells are dispersed into a homogeneous suspension. Add 15 mg of muramidase and incubate the suspension for 20 min in a shaking water bath at 37°C, then stir in 6 ml of SDS and place in a water bath at 60°C for 10 min. Cool on ice to room temperature and decant into a 250 ml stoppered bottle. Add 7 ml of sodium chlorate solution with stirring and vigorously shake for 15 min with 40 ml of chloroform/isoamyl alcohol.

First ethanol precipitation   Centrifuge the resulting emulsion for 5 min at 10 000 g in stoppered opaque centrifuge tubes. Transfer the upper aqueous phase to a 250 ml beaker and gently layer 80 ml of ethanol over the solution by pouring slowly down the side of the beaker. Mix the phases gently with a glass rod and wind out the DNA fibres on the rod. Squeeze as much liquid as possible from the spooled mass by pressing it against the side of the beaker, blot with filter paper to remove any residual ethanol.

RNase and second ethanol precipitation   Dissolve the DNA in 10 ml of SSC in a 50 ml conical flask, then add RNase solution to a final concentration of 100 µg/ml and incubate at 37°C for 30 min, with occasional shaking.

Stop the reaction by adding 2 ml of sodium chlorate solution, followed by 15 ml of chloroform/isoamyl alcohol and, after shaking slowly for 10 min, centrifuge as above.

Precipitate the DNA from the upper aqueous layer with 30 ml ethanol, spool out the DNA and redissolve it in 10 ml of SSC. Store at 4°C.

Remove 0.5 ml for the assay of DNA, 0.5 ml for the assay of RNA and 0.1 ml for the determination of the absorption spectrum.

Isopropanol precipitation   Add 1 ml of sodium acetate to the remaining 9.3 ml of DNA, mix thoroughly and slowly drip in 6 ml of isopropanol. Spool out the DNA fibres while gradually mixing the phases together. If a gel-like precipitate forms, add a further 1–2 ml of isopropanol. Press the residual solvent from the DNA, blot with filter paper and dissolve in 5 ml of SSC.

Assay for nucleic acids   Assay suitable aliquots of the second ethanol precipitate and the purified DNA for nucleic acids using the diphenylamine assay for DNA and the orcinol assay for RNA. In both cases, adjust the sample volume to 2 ml with SSC then proceed as described in Chapter 11.

UV absorption of DNA   Obtain a UV absorption spectrum for the sample of DNA and calculate the concentration of the nucleic acid by assuming that 1 mg/ml of DNA has an extinction of 20.0 at 260 nm in a 1 cm cuvette.

Compare the shape of your spectra with published data and comment on the purity of your preparation. The extinction values at 260 nm and 280 nm can be used to detect protein. The E_{260}/E_{280} ratio should be between 1.8 and 1.9 for pure DNA and a value less than 1.75 indicates significant protein contamination.

Add 0.2 ml of 6 mol/litre NaOH to the DNA samples in the cuvettes and repeat the UV spectral scan. Wind the recorder back so that the spectrum is superimposed on the previous one and comment on the results.

Quality of DNA   Evidence for *degradation* by shearing of the DNA can be obtained by determination of the molecular weight by viscometry (Exp. 11.7) and comparison with published values. Evidence for *denaturation* (strand separation) can be obtained from hyperchromicity measurements (Exp. 11.6).

Solubility of DNA   Difficulty may sometimes be encountered in dissolving DNA precipitates but this can be partly avoided by pressing out all residual ethanol on the sides of the beaker. Warming to 50°C may also hasten solubility. If the precipitate still does not dissolve completely, spin on the bench centrifuge for 5 min, remove the supernatant into a vial and dissolve the residue in 1–2 ml of dilute SSC. Finally, bulk the solutions making a note of the total volume.

Comments   During the extraction procedure note the point at which there is a sudden change in viscosity. What does this denote?

Comment on the physical form of the precipitates obtained with ethanol and isopropanol.

Experiment 16.10   The isolation and characterization of eukaryotic mRNA

PRINCIPLE

Phenol extraction procedures yield a mixture of all types of RNA including mRNA, although this is a minor component compared to the large amounts of rRNA. Further purification of eukaryotic mRNA depends on the presence of poly(A) tails on the 3′ ends of mRMA and involves passing the crude RNA through a small column of oligo (dT) cellulose. Only mRNA with a complementary poly(A) base sequence will bind while tRNA, rRNA and proteins should pass through. The mRNA fraction can then be eluted by lowering the ionic strength. In effect, therefore, this is a form of affinity chromatography. The oligo (dT) chains, which have been chemically linked to the cellulose support are about 18 nucleotides long.

Rather than extracting total RNA directly from cells it is often more convenient to prepare a crude polyribosome fraction by centrifugation and then extract this with phenol/chloroform to give a crude RNA fraction.

All reagents and apparatus should be sterilized before use.

Note   Considerable care is needed to minimize the action of RNase. This enzyme is present on skin and is often found contaminating laboratory glassware and reagents. Most reagents and apparatus should therefore be autoclaved and it is necessary to wear plastic gloves.

MATERIALS   10

1. Reticulocyte lysate (Exp. 16.7)
2. Hypertonic sucrose (1 mol/litre sucrose, 25 mmol/litre KCl,   100 ml
    1 mmol/litre $MgCl_2$)
3. Alkaline buffer (0.1 mol/litre tris–HCl, 0.1 mol/litre NaCl,   1 litre
    1 mmol/litre EDTA, pH 9.0)
4. UV spectrophotometers   5
5. Sodium docecyl sulphate (25 per cent w/v)   50 ml
6. Phenolic extraction medium. (Redistill the phenol, saturate with   1 litre
    the alkaline buffer and mix 500 ml with 500 ml of chloroform and
    10 ml of isoamyl alcohol)
7. Separation funnels (500 ml)   5
8. Sodium acetate (20 per cent w/v, pH 5.5)   100 ml
9. Ethanol   2 litres
10. Oligo (dT) cellulose   5 g

11. NTS buffer (0.5 mol/litre NaCl, 10 mmol/litre tris–HCl, 0.1 per   1 litre
    cent w/v SDS, pH 7.4)
12. TS buffer (10 mmol/litre tris–HCl, 0.1 per cent w/v SDS, pH 7.4)   1 litre
13. Sterile syringes (2 ml)   5
14. Whatman GF/B discs   10
15. Nylon cloth   —
16. NaOH (0.1 mol/litre)   100 ml
17. Polyadenylic acid (1 mg/ml in NTS buffer)   2 ml
18. Polyuridylic acid (1 mg/ml in NTS buffer)   2 ml
19. tRNA   2 mg
20. Refrigerated ultracentrifuges   5
21. Sterile water   —
22. Tris buffer (10 mmol/litre tris–HCl, pH 7.5)   250 ml
23. Deep freeze at $-20°C$ 1
24. Deep freeze at $-70°C$ 1

METHOD

Isolation of polyribosomal RNA   Collect reticulocytes from a rabbit to give about 30 ml of packed cells and lyse with two volumes of distilled water. Centrifuge the lysate at 20 000 g for 10 min then layer the supernatant over 1/10 volume of hypertonic sucrose and spin at 40 000 g for 3 h. Resuspend the pellet in alkaline buffer, determine the extinction on a small aliquot and adjust the volume of buffer to give an extinction of 20 per ml at 260 nm. Add SDS solution to give a final concentration of 1 per cent w/v of the detergent, then add an equal volume of the phenol extraction medium. Shake for 10 min at room temperature, cool on ice and spin at 3000 g for 10 min.

Re-extract the top aqueous phase with an equal volume of the phenol mixture and retain the upper aqueous phase. Re-extract the first phenol phase and the interphase with an equal volume of buffer and keep the aqueous phase. Combine the aqueous phases obtained from the re-extractions and repeat if the final interphase is not clean (i.e., mRNA still trapped with protein).

Add 20 per cent w/v sodium acetate to give a final concentration of 2 per cent then add 2 volumes of ethanol at $-20°C$ and leave for at least 2 h but preferably overnight. Centrifuge the RNA at 5000 g for 40–60 min and wash the precipitate 2–3 times with 75 per cent v/v ethanol. Dry under vacuum. Resuspend in NTS buffer to give an extinction at 260 nm of 50–100 per ml and freeze at $-70°C$ in small aliquots.

Column preparations   Suspend 0.2 g of oligo (dT) cellulose in the TS buffer and pack the material into a 2 ml sterile syringe the base of which contains a Whatman GF/B disc over nylon cloth. Wash the column with several bed volumes of 0.1 mol/litre NaOH, then remove the NaOH by washing thoroughly with water. Equilibrate the column with 10 volumes of the NTS buffer and check that the pH of the eluate is in the region of 7.4–8.0. The background extinction should be less than 0.05 at 260 nm.

Column performance   Prepare two columns and apply 100 µg of poly (A) to one and 100 µg of poly (U) to the other. Elute with NTS buffer at a flow rate of 1 ml/min and collect 2 ml fractions until the E_{260} returns to the baseline. Continue the elution with the TS buffer and again monitor until the E_{260} is constant. Calculate the recovery of poly(A) and poly(U) from the columns and account for the difference in the elution profiles.

Column regeneration   Wash the column with 0.1 mol/litre NaOH followed by NTS buffer until the pH approaches neutrality.

Preparation of mRNA using the oligo (dT) columns   The crude RNA preparation should now be applied to the column. Up to 100 extinction units may be applied to a 0.2 g column but ideally apply 20 extinction units in 1 ml. Collect the eluate and re-apply to the column. Then wash the column with NTS buffer until the E_{260} reaches baseline (approx. 10 ml). Determine the E_{260} and volume of the total eluate. Elute the bound RNA with TS buffer (approx. 5–10 ml), collecting 2 ml fractions on ice. Determine the E_{260} of each fraction and estimate the yield of mRNA as a per cent of the total applied.

Precipitation of mRNA   Adjust the ionic strength of the solution by adding sodium acetate to give a final concentration of 2 per cent w/v. Ensure that the RNA concentration is at least 1 unit per ml by the addition of tRNA. This acts as a carrier to aid coprecipitation in the next step. Precipitate the RNA by adding 2 volumes of ethanol and stand at $-20°C$ for at least 2 h. Centrifuge at 10 000 g for 20 min, then dissolve the pellet in sterile water or 10 mmol/litre tris buffer to give a concentration of 20 units per ml and store frozen at $-70°C$.

Comments   Discuss the elution profiles obtained after application of the RNA sample to the oligo (dT) column. What proportion of RNA came off in the two buffers?

How would you show that mRNA contains a 3′ poly(A) tail? Is this present on all species of mRNA in both prokaryotes and eukaryotes?

Why are conditions of low ionic strength used to elute mRNA from the column?

What are the relative proportions of RNA found in cells?

Experiment 16.11   Translation of mRNA from reticulocyte lysates

PRINCIPLE

The lysate is first treated with RNase to eliminate endogenous mRNA and the isolated mRNA is added to see if the treated lysate has the ability to translate it.

As in the previous experiment all reagents should be sterile and the apparatus autoclaved.

MATERIALS

1. Purified mRNA (Exp. 16.10)
2. Haemin solution (Exp. 16.8)

$$\frac{10}{-}$$
$-$

3. CaCl$_2$ (0.1 mol/litre)	10 ml
4. RNase (15 000 units/ml, stored at $-20°C$)	100 μl
5. Shaking water baths at 20°C	5
6. EGTA (0.2 mol/litre, pH 7.4)	5 ml
7. tRNA from calf liver	0.5 mg
8. Reagents for Exp. 16.7	—

METHOD

Preparation of nuclease-treated lysate   Mix 40 μl of haemin with 10 μl of CaCl$_2$, equilibrate at 20°C then add 10 μl of micrococcal RNase and incubate by shaking at 20°C.

Translation of reticulocyte mRNA   Prepare a translation mixture as for the untreated lysate but add mRNA as template. Approximately 15 μg of *pure* mRNA would be ideal but up to 2 mg of RNA ($E_{260} = 50$) may have to be added if the mRNA is impure. *Note:* 40 μg/ml RNA gives an absorbance of 1.0 at 260 nm.
    Determine the rate of protein synthesis using this system (Exp. 16.7).

Further reading

Alberts, B., Bray, D., Lewis, J., Roff, M., Roberts, K. and Watson, J. D., *Molecular Biology of the Cell*. Garland Publishing, New York, 1983.

Glover, D. M., *Gene Cloning*. Chapman and Hall, London, 1984.

Levine, L., *Biology of the Gene*, 3rd edn. C. V. Mosby, St Louis, 1980.

Macleod, A. and Sikora, K., *Molecular Biology and Human Disease*. Blackwell Scientific Publications, Oxford, 1984.

Rees, A. R. and Sternberg, M. J. E., *From Cells to Atoms: An Illustrated Introduction to Molecular Biology*. Blackwell Scientific Publications, Oxford, 1984.

Appendix

Chapter 2   Accuracy

Answers to calculations

1. 10.8 g of glucose are needed per 100 ml for a 0.06 mol/litre solution.
2. (a) NaCl 0.15 mol/litre = 0.15 mmol/ml = 150 µmol/ml.
   (b) Fructose 12 mmol/litre = 12 µmol/ml.
   (c) ATP 200 µmol/litre = 200 nmol/ml = 0.2 µmol/ml.
   (d) Urea 300 g/litre = 300/60 = 5 mol/litre = 5 mmol/ml.
3. Glycine in 10 ml = 15 mg.
4. Fasting blood glucose = 90 mg/100 ml.
5. Plasma protein = 6.5 mg/ml.
6. Enzyme activity = 45 nmol/min.

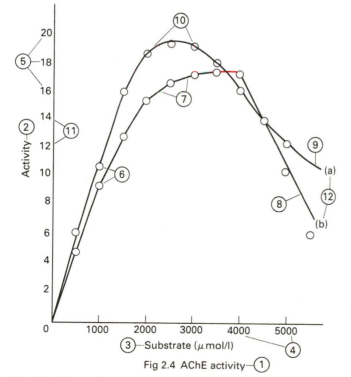

Fig. A1   Errors in Fig. 2.4

Errors on the graph (Fig. 2.4)

1. Insufficient information in the title. Do not use abbreviations unless they are first defined.
2. No units given for the activity.
3. What is the substrate?
4. Too many digits in abscissa, use mmol/litre.
5. Too many numbers in the ordinate.
6. Same symbols used for both curves.
7. Join up the points of curve (b) with a smooth curve not a series of straight lines.
8. Curve should follow experimental points.
9. Curve should not extend beyond the last experimentally determined point.
10. Lines should not run through symbols.
11. No scale marks given.
12. What do the curves (a) and (b) represent?

Chapter 3   pH and buffer solutions

Answers to calculations

1. (a) pH 1; (b) pH 13; (c) pH 2; (d) pH 1.
2. $pK_1 = 4.19$, $pK_2 = 5.57$.
3. H^+ at pH 7.4 $= 3.80 \times 10^{-8}$ mol/litre.
   In acidosis with an H^+ of 7.6×10^{-8}, the pH is 7.12.
4. $pK_a = 4.21$, $K_a = 6.17 \times 10^{-5}$.
5. Total glycine is 0.1 mol/litre.
   For pH 2.0, HCl needed $= 0.0715$ mol/litre.
   For pH 3.0, HCl needed $= 0.02$ mol/litre.

Buffer solutions

ACETATE BUFFERS (0.2 mol/litre)
Place x ml of 0.2 mol/litre acetic acid in a 100 ml volumetric flask and make up to the mark with 0.2 mol/litre sodium acetate.

x ml	92	88	83	76	66	55	43	32	22	17	7
pH	3.6	3.8	4.0	4.2	4.4	4.6	4.8	5.0	5.2	5.4	5.6

CARBONATE/BICARBONATE BUFFERS (0.1 mol/litre)
Place x ml of 0.1 mol/litre sodium bicarbonate into a 100 ml volumetric flask and make up to the mark with 0.1 mol/litre sodium carbonate.

x ml	93	88	82	73	61	49	38	27	18	10	5.5
pH	9.0	9.2	9.4	9.6	9.8	10.0	10.2	10.4	10.6	10.8	11.0

CITRIC ACID/SODIUM CITRATE BUFFERS (0.05 mol/litre)
Place x ml of 0.05 mol/litre citric acid in a volumetric flask and make up to the mark with 0.05 mol/litre trisodium citrate.

x ml	91	86	80	75	70	65	60	55
pH	3.0	3.2	3.4	3.6	3.8	4.0	4.2	4.4

x ml	50	44	39	34	29	24	19	14
pH	4.6	4.8	5.0	5.2	5.4	5.6	5.8	6.0

PHOSPHATE BUFFERS (0.1 mol/litre)

Sodium phosphate buffers   Add x ml of 0.2 mol/litre sodium hydroxide to 50 ml of 0.02 mol/litre sodium dihydrogen phosphate and dilute to 100 ml.

Potassium phosphate buffers   Add x ml of 0.2 mol/litre potassium hydroxide to 50 ml of 0.2 mol/litre potassium dihydrogen phosphate and dilute to 100 ml.

x ml	3.5	5.8	9.1	13	18	24	30	35	40	43	45	47
pH	5.8	6.0	6.2	6.4	6.6	6.8	7.0	7.2	7.4	7.6	7.8	8.0

TRIS(HYDROXYMETHYL)AMINO METHANE/HCL (0.1 mol/litre)

Add x ml of 0.2 mol/litre hydrochloric acid to 50 ml of 0.2 mol/litre tris and make up to 100 ml.

x ml	43	41	39	34	29	24	18	13	9.5	6.0
pH	7.2	7.4	7.6	7.8	8.0	8.2	8.4	8.6	8.8	9.0

MISCELLANEOUS BUFFERS

Barbitone (0.07 mol/litre, pH 8.6)   Dissolve 2.58 g of diethylbarbituric acid and 14.42 g of sodium diethylbarbiturate in water and make up to 1 litre.

Cacodylate (0.1 mol/litre, pH 6.0)   Add 29 ml of 0.2 mol/litre HCl to 50 ml of sodium cacodylate and make up to 100 ml with water.

Cacodylate (0.1 mol/litre, pH 6.5)   Add 16 ml of 0.2 mol/litre HCl to 50 ml of sodium cacodylate and make up to 100 ml with water.

Phthalate (0.1 mol/litre, pH 6.0)   Add 43 ml of 0.1 mol/litre NaOH to 50 ml of 0.2 mol/litre potassium hydrogen phthalate.

Chapter 7   Spectrophotometry

Answers to calculations

1. (a)   Absorbance = 0.154.
   (b)   Transmittance = 34.5 per cent.
   (c)   Absorbance = 0.77.
2. Glycogen concentration = 0.6 g/100 ml.

3. (a) Molar extinction coefficient $= 5 \times 10^3$ litres mol^{-1} cm^{-1}.
   (b) Transmittance $= 89.1$ per cent.
4. Tryptophan $= 0.1$ mmol/litre.
   Tyrosine $= 0.1$ mmol/litre.
5. Plasma albumin $= 21.6$ mg/ml.
6. NADH $= 50$ μmol/litre.
   NAD^+ $= 20$ μmol/litre.
7. Using $F = KI_0(1 - e^{-kcl})$, then the fluorimeter reading for 40 μmol/litre $= 19$.

Chapter 8   Amino acids and proteins

Free amino and carboxyl end groups of proteins

Protein		$-NH_2$	$-COOH$
Haemoglobin α chain		Valine	Arginine
	β chain	Valine	Histidine
Insulin	A chain	Glycine	Asparagine
	B chain	Phenylalanine	Alanine
Muramidase		Lysine	Leucine
Ribonuclease		Lysine	Valine

Standard bovine serum albumin

Most commercial preparations of bovine serum albumin (BSA) contain fat which should be removed before using the protein as a standard. Preparing a solution of BSA from the defatted material can also be a problem and the hints given below should be carefully followed to avoid difficulties.

Defatting of BSA   Add 10 g of BSA to 100 ml of a mixture of 95 per cent v/v ethanol and ether (3:1) and stir for 15 min at 0°C. Decant the solvent and repeat the washing at 0°C. Filter the albumin and wash in ether until dry.

Preparation of standard BSA (5 mg/ml)   Accurately weigh 0.5 g of the defatted BSA and add this to about 80 ml of water in a beaker. Do not stir or shake but leave to stand until the albumin has dissolved. Gently swirl the solution to mix it then transfer the albumin to a 100 ml volumetric flask. Rinse out the beaker several times with water and add the washes to the volumetric flask. Carefully invert the flask several times to mix the contents and make up to the mark with water.

The secret of the whole process is not to shake or stir vigorously as this leads to denaturation and severe frothing which then makes it virtually impossible to prepare an accurate solution.

Storage of the standard BSA   Bovine serum albumin is not a cheap reagent and should not be wasted. In many cases an accurate solution needs to be prepared of which only a small volume is needed. It is therefore recommended that the 100 ml be dispensed into 5 ml aliquots and stored in the deep freeze until required. Once a tube of BSA solution is thawed for use, any excess remaining should be discarded and not frozen again.

Chapter 10   Lipids

Properties of some natural and synthetic lipids

Oil or fat	Acid value (mg KOH/g)	Iodine value (g I_2/100 g)	Saponification number (mg KOH/g)
Butter	0.5–30	25–40	210–230
Castor oil	0.2–4	80–90	175–187
Coconut oil	2.5–6	7–10	254–262
Cod liver oil	0.5–5	140–180	180–190
Corn oil	1.0–2	104–128	187–195
Cow's milk		26–45	216–235
Glyceryl tributyrate	0	0	557
Glyceryl trioleate	0	86	190
Glyceryl tristearate	0	0	189
Halibut liver oil	1	120–135	170–180
Linseed oil	1.0–4	170–195	188–195
Olive oil	0.2–3	80–90	190–195

Cholesterol, m.p. 149°C

Chapter 12 Enzymes

Data on enzymes

Enzyme	Substrate	Optimum pH	Michaelis constant (mmol/litre)
Acetylcholinesterase (human erythrocytes)	Acetylcholine	8.0	0.075
Acid phosphatase (human erythrocytes)	β-Glycerophosphate	5.0	0.20
Alanine aminotransferase (rat liver)	L-Alanine 2-Oxoglutarate	8.0	42.0 1.1
Alkaline phosphatase (human kidney)	p-Nitrophenyl- phosphate	10.0	0.56
α-Amylase (human saliva)	Starch	6.9	0.6 mg/ml
β-Amylase (potato)	Amylose	4.8	0.073
Cholinesterase (human serum)	Benzoyl choline	7.5	0.004
β-Fructofuranosidase (yeast)	Sucrose	4.8	9.1
Glucose oxidase (*Penicillium notatum*)	Glucose	5.6	9.6
Glutamate dehydrogenase (ox liver)	Glutamate NADP	8.0	1.8 0.047
Lactate dehydrogenase (rabbit muscle)	Pyruvate	7.8	0.42
Trypsin (pancreas)	Casein	7.8	21 mg/ml

Note   pH optima and K_m values are not absolute constants but depend on the temperature, substrate and source of enzyme.

Index